Portrait of a Prospector

Edward Schieffelin, 1877. "He was the queerest specimen of humanity ever seen . . . with black curly hair that hung several inches below his shoulders. His long untrimmed beard was a mass of unkept knots and mats. . . . Although only 27 years of age, he looked at least 40." "Description of Ed Schieffelin at age twenty-seven" in McClintock, *Arizona Prehistoric, Aboriginal, Pioneer, Modern*, 3:412. *Photo courtesy Arizona Historical Society.*

Portrait OF A Prospector

Edward Schieffelin's Own Story

Edited by R. Bruce Craig

UNIVERSITY OF OKLAHOMA PRESS : NORMAN

Library of Congress Cataloging-in-Publication Data

Name: Schieffelin, Edward, 1847–1897, author. | Craig, R. Bruce, editor.
Title: Portrait of a prospector : Edward Schieffelin's own story / edited by R. Bruce Craig.
Other titles: Edward Schieffelin's own story
Description: First edition. | Norman, OK : University of Oklahoma Press, [2017] |
    Includes bibliographical references and index.
Identifiers: LCCN 2017010496 | ISBN 978-0-8061-5773-3 (pbk. : alk. paper)
Subjects: LCSH: Schieffelin, Edward, 1847–1897—Diaries. | Tombstone Region
    (Ariz.)—History | Tombstone Region (Ariz.)—Biography—Pictorial works. |
Mines and mineral resources—Arizona—Tombstone Region—History. | Pioneers—
    Arizona—Tombstone Region—Biography. | Miners—Arizona—Tombstone
    Region—Biography. | Tombstone Region (Ariz.)—Biography.
Classification: LCC F819.T6 S327 2017 | DDC 979.1/53—dc23
LC record available at https://lccn.loc.gov/2017010496

The paper in this book meets the guidelines for permanence and durability of the Committee on Production Guidelines for Book Longevity of the Council on Library Resources, Inc. ∞

Copyright © 2017 by the University of Oklahoma Press, Norman, Publishing Division of the University. Manufactured in the U.S.A.

All rights reserved. No part of this publication may be reproduced, stored in a retrieval system, or transmitted, in any form or by any means, electronic, mechanical, photocopying, recording, or otherwise—except as permitted under Section 107 or 108 of the United States Copyright Act—without the prior written permission of the University of Oklahoma Press. To request permission to reproduce selections from this book, write to Permissions, University of Oklahoma Press, 2800 Venture Drive, Norman, OK 73069, or email rights.oupress@ou.edu.

Contents

List of Illustrations	*vii*
Acknowledgments	*ix*
Chronology of the Life of Edward Schieffelin	*xi*
Introduction	*3*
1. Early Life, 1847–1866	*18*
2. "Grubstakes" Early Wanderings, 1866–1872	*26*
3. Up the Colorado, 1872	*34*
4. Southwestern Wanderings, 1872–1877	*44*
5. The Discovery of Tombstone, 1878	*55*
6. Alaskan Adventures, 1882–1883	*70*
7. Ever a Prospector, 1883–1897	*98*
Glossary	*103*
Bibliography	*105*
Index	*111*

Illustrations

FIGURES

Edward Schieffelin, 1877	*frontispiece*
Schieffelin Hall, 2016	4
First page of Schieffelin's memoirs	13
Placer mining woodcut, ca. 1860	19
Print depicting the Mountain Meadows Massacre	29
Tombstone, Arizona, 1880s	67
Alaska prospecting party, 1882	72
The *New Racket*, 1882	75
Schieffelin's Alaskan dog team, 1882	82
Nuklukayet Trading Station, 1882	84
Schieffelin and wife, M. E. Brown, 1883	95
Schieffelin grave and monument, ca. 1900	100

MAP

Edward Schieffelin's West	2

Acknowledgments

A BOOK LIKE THIS draws upon the encouragement and support of many. I am indebted to a great number of individuals who over the years prodded me to compile Edward Schieffelin's memoirs and writings. A special note of gratitude is due to the late David Lavender whose encouragement to an undergraduate history student remains as clear in my mind today as when first given decades ago. Also, I owe a hearty thanks to the late Dr. Charles C. McLaughlin of the American University. His initial editorial suggestions and thoughtful comments provided a needed critique that served to sharpen my own thinking about many aspects of Schieffelin's life when I embarked on this project back in graduate school.

I am indebted to the Bancroft Library at the University of California–Berkeley, to the Western History Collections of the University of Oklahoma Libraries, and to the Arizona Historical Society for granting permission to reproduce manuscript and photographic materials in their respective collections. Thanks also to the Huntington Library in San Marino, California, for permission to reprint the photographs of the Schieffelin Alaskan trip taken by Charles O. Farciot. Dawn Hayes, interlibrary loan specialist at the University of Prince Edward Island's Robertson Library, deserves kudos for tracking down a fair number of rather obscure books and articles needed to round out the manuscript's footnotes. Keith Davis of Goose Flats Graphics and Publishing was so kind as to provide the photograph of Tombstone's Schieffelin Hall, and cartographer Tom Jonas tackled the job of preparing a map locating a list of historic places (some of which no longer exist) that help to document Schieffelin's various forays into the southwestern desert.

A special thanks to the editorial and production team at the University of Oklahoma Press, including Director B. Byron Price, Copy Editor Ariane C. Smith, Acquisitions Editor J. Kent Calder, Marketing Assistant Amy Hernandez,

ACKNOWLEDGMENTS

and especially to Manuscript Editor Emily Schuster whose suggestions brought the text into its final form. Thanks also to Daphne Davey of Rural Dynamics who contributed mightily to this project, not only by performing the initial editing of the manuscript but also by assembling the index in near record-breaking time. And to my wife, Patricia, whose experience at the Smithsonian Institution and the Manuscript Division of the Library of Congress caused her sharp editorial pencil never to dull. All deserve special thanks for support and patience.

Chronology of the Life of Edward Schieffelin

October 8, 1847. Born in Tioga County, Pennsylvania.

1852. Father Clinton Schieffelin goes to California, then Oregon. Claims land along the Rogue River near Jewett's Ferry, Oregon.

Fall 1856. Travels with mother and brothers from Pennsylvania to California to meet father.

February 1857. Reunited with father in California. Family travels to new home in Oregon.

1860–61. At age twelve, runs away from home to search for gold during the Salmon River "excitement." Brought home by a sympathetic neighbor.

1866. First prospecting trip in southwestern Oregon with brother Albert (Al).

1866–71. Initial wanderings in California and Nevada.

1871. Lured to Utah by Salt River "excitement."

March 1872. Passes through Mountain Meadows massacre area.

Spring 1872. Barely escapes death at the hands of renegade Indians.

March–April 1872. Travels up the Colorado River into the Grand Canyon with four companions.

Summer 1872. Wanders and prospects in the Southwest, encountering Indians; narrowly escapes a flash flood in a box canyon.

1874–75. Returns home to Oregon after six years of unsuccessful prospecting.

CHRONOLOGY

November 1875. Borrows one hundred dollars from father and departs for Arizona.

Summer–Fall 1877. Prospects in Hualapai country and joins William T. Griffith and Albert "Alvah" Smith. Assists in the assessment work at the Brunckow Mine. Begins prospecting in Tombstone region and becomes convinced he has located "rich diggins."

August 1877. Without provisions and clothes, joins up with W. H. Sampson and prospects jointly in the Tombstone region for about a month.

September 1877–January 1878. Seeks out brother Albert in the Globe mining district. Joins up with Albert and Richard Gird, who confirms that Schieffelin's gold-bearing quartz is high-grade ore.

January 1878. Forms partnership with brothers and Gird.

February–March 1878. Partnership stakes claims in Tombstone district.

October 1878. Incorporates Schieffelin/Gird partnership. Starts to develop mining properties with funding from eastern investors.

1879–80. Corbin brothers invest in mines.

March 1880. With brothers, sells all interests in mines. Returns home to parents' farm in Oregon.

1880–82. Family relocates to Los Angeles. Buys home for aging parents.

March 1882. Makes trip to Juneau, Alaska, with brother Eff. Returns home to make arrangements for a major expedition up the Yukon River.

June 13, 1882. Schieffelin expedition departs San Francisco for Alaska.

1882–83. Alaskan adventures.

April 1883. Father dies of accidental gunshot.

Fall 1883. Returns to San Francisco wanting "no more of Alaska." Marries Mary Elizabeth Brown in La Junta, Colorado, and winters briefly in Salt Lake City.

CHRONOLOGY

Spring 1884. Buys home in Alameda, California. Interviewed by historian Hubert Howe Bancroft.

Fall 1884. Makes an unsuccessful prospecting trip into the Papago Indian country.

1885–95. The "lost years." Makes several additional prospecting trips and a grand tour of the East. Treated as a celebrity, visits New York, Washington, D.C., Philadelphia, and other eastern cities.

1896. Returns "home" to Oregon to prospect.

September 1896. Briefly returns home to make out his last will and testament.

1896–97. Returns to Oregon. Legend has it that he discovers "rich diggins" in vicinity of Canyonville.

May 12, 1897. Dies of an apparent heart attack in small cabin twenty miles from Canyonville. Whereabouts of Red Blanket Mine is lost.

May 23, 1897. Family conducts funeral in Tombstone. Remains interred at location of first campsite in Tombstone district.

Portrait of a Prospector

Edward Schieffelin's West. *Map by Tom Jonas. Copyright © 2017, University of Oklahoma Press.*

→ ←

Introduction

DURING THE LATE NINETEENTH CENTURY, a larger-than-life-sized portrait of a bearded prospector greeted thousands as they passed through the halls of the California State Mining Bureau in San Francisco's old Ferry Building. A prime example of American individualism, the man in this gilt-framed portrait—with a pick and shovel in hand and a canteen and a gun draped across his leg—reinforced a popular American myth: that merely with determination, persistence, and a few bare necessities, anyone could achieve their dreams in the American West.

The striking figure was Edward Lawrence Schieffelin: Indian scout, stagecoach driver, Alaskan explorer, discoverer of Arizona's rich Tombstone silver mines, and, above all else, the West's most famous prospector.

This book is a portrait of a prospector—not of a typical one, mind you, but a successful one. In the late nineteenth century he was known to millions, but today few know of him. Students in a class on the American West may run across his name in a western history textbook. Tourists visiting Tombstone, the town "too tough to die," cannot help but see his name when they set their eyes on Schieffelin Hall,[1] the one opulent building in town named after the founding father of that community that came to epitomize the "wild West." Backpackers studying U.S. Geological Survey maps in anticipation of journeying into the Alaska wilderness may recognize the name Schieffelin, which is assigned to a small tributary of the mighty Yukon River. And a few old-timers

---

1. Built in 1881 and located at the corner of Fourth and Fremont Streets, Schieffelin Hall was the largest adobe building in the United States in its day—two stories high and 130 feet long. Alfred Schieffelin, who felt that Tombstone needed a first-rate theater, had it built in his brother's honor. The 700-seat edifice became the social center of the town. Melodramas, musicals, plays, and, in 1897, Ed Schieffelin's funeral—all took place there. See Way, *The Tombstone Story*, 17; Barnes, *Arizona Place Names*, 116–17.

Schieffelin Hall, the social center of Tombstone. *Courtesy Keith Davis, Goose Flats Graphics.*

living near Canyonville, Oregon, may remember his name in association with legendary stories handed down by their fathers and grandfathers—stories of a brief gold rush touched off in that country when, in the second week of May 1897, evidence of a "rich prospect" was discovered near the body of the dead prospector.

Today, when we look at faded photographs of this once-famous, full-bearded pioneer, we see the epitome of what historians of the American West, such as Frederick Jackson Turner and Ray Billington, have characterized as one of the "frontier types."[2] Schieffelin's encounter with the West, though, was unique. Unlike so many of his less fortunate prospecting compatriots, his perseverance ultimately paid off—he died a wealthy man. Though his successes were atypical of most prospectors, his adventures, both before and after his "lucky strike" at

---

2. Turner, *The Frontier in American History*; Billington, *America's Frontier Heritage*; see also Rundell, "Concepts of the 'Frontier' and the 'West,'" 13–41. Though critical of Turner's frontier thesis, historians of the "new western history" nevertheless recognize the existence of various frontier "types." See Limerick, *Trails*, 120.

INTRODUCTION

Tombstone, indeed did reflect the experiences of thousands of others. With its allure of riches and its prospect for personal gain, the molding effect of this frontier experience helped shape the perception of a distinctive "American" civilization. Ed Schieffelin's story helps to explain what America was—and perhaps is.

Ed Schieffelin: A Prospector's Life

Edward Lawrence Schieffelin was born in the coal-mining region of Tioga County, Pennsylvania, on October 8, 1847. Like other pioneers lured by the 1848–49 discovery of gold in California, Schieffelin's father, Clinton, left the comforts of the "civilized" East in the 1850s to make his fortune in the gold camps of the West. His prospecting efforts, however, did not pan out. Both Clinton and his brother Jacob ultimately gave up their dream. Purchasing property near Jewett's Ferry, they settled down to a life of farming in the Rogue River region in southwestern Oregon. In 1856 Clinton sent for his wife, the former Jane L. Walker, and their young sons to join him in Oregon. Though the brothers continued to prospect part-time, their efforts focused largely on breeding cattle, harvesting grain, and raising a family.[3]

Along the sparkling banks of the Rogue River, among the rolling wheat fields and neatly planted orchards, the young Ed Schieffelin first experienced the excitement of prospecting. It is not hard to picture a ten-year-old boy crouched down along a river bank, swishing water and gravel back and forth in his pan. For several hours he worked to collect half a teaspoon of yellow flakes—what he called his "first prospect." Carefully, he put the ore in a tiny bag before dashing home. There, breathless, he shared news of his big discovery with his family.

After carefully examining his nephew's pickings, Ed's uncle chuckled and told him that his bag of gold was nothing more than worthless mica. But the disappointing news did not dampen the boy's enthusiasm. He was hooked. From that day forward, even after he had made all the money he could possibly spend in a lifetime, Schieffelin continued to try to replicate that first feeling

---

3. For Schieffelin family history, see "The Schieffelin Family Collection," Beinecke Rare Book and Manuscript Library, Yale University, New Haven Conn., accessed January 5, 2016, http://hdl.handle.net/10079/fa/beinecke.schieff.

of elation that came from that boyhood discovery. As he described it, that experience "infected" him with gold fever. "If I had a fortune when I was 22," he later wrote, "I suppose I'd not keep it long; for now I think of it I can't see why I should . . . I like the excitement for being right up against the earth, trying to coax the gold away to scatter it."[4]

As a part-time prospector himself, Clinton Schieffelin encouraged all his sons to prospect. However, his second son's preoccupation with gold eventually caused his parents some consternation. At age twelve, for example, Ed ran away from home, hoping to join the Salmon River "excitement" in Idaho, only to be dragged back home after some sixty miles by an understanding neighbor. Schieffelin's parents could do little to control Ed's urges to prospect (or, for that matter, that of his brothers, Albert Eugene and Effingham Lawrence). If it had not been for his parents' objections, Ed probably would have left home at age seventeen to join the mining frenzy in Montana. Though his parents convinced him to stay closer to home and instead work the placer mines located a mere five miles away, nothing could contain young Ed's dreams. He remained close to home only for about a month. Shortly before his eighteenth birthday, with money he had saved securely fastened in his knapsack, he struck out on his own.

Like most young men who had the gleam of gold in their eyes, Schieffelin embarked on a series of unproductive prospecting escapades. And like thousands of others, he jumped at the opportunity to move on at the first word of a new discovery of "rich diggins" in some far-off country. "My long-continued ill successes," he later reflected, "was owing I believe to following other men's footsteps and follow-up excitements."

And follow excitements he did. He prospected for silver ore on California's Humboldt River; he tried his luck unsuccessfully at gold along the Owens River; he packed up his pick and shovels and joined the "Salt River excitement"—all without success. Over the years he traveled the West extensively, leaving tracks in the Death Valley of California, as well as Nevada, Colorado, New Mexico, Utah, Idaho, and Arizona. There, perched on a hastily constructed boat flying a little flag with the words "Good Hope" waving in the breeze, he and several companions embarked on yet another unsuccessful trip—this time to find their El Dorado in the "grandest of all canons," the Grand Canyon.

---

4. Schieffelin as quoted in Erwin, *The Southwest of John H. Slaughter*, 181.

INTRODUCTION

Finally, in what at first appeared to be among the most unlikely of all places, in a desolate area in the Arizona desert surrounded by the Whetstone, Mule, Burro, Huachuca, and Dragoon Mountains—and under the watchful eye of Apache Indians—Schieffelin finally hit pay dirt.

The story of Ed Schieffelin's Tombstone discovery remains a standard of western lore.[5] In January 1877, with two mules, a brand new prospecting outfit, and thirty dollars in his pocket, Schieffelin struck out to explore the "Hualapai country" in the vicinity of present-day Mohave County, Arizona. He had joined the surge of ever-hopeful prospectors in the search for "rich prospects" that had been reportedly found in the southeastern part of the territory. Schieffelin concentrated his efforts in what were considered dangerous hills infested with "troublesome Indians" east of the San Pedro River. At Camp Huachuca, at the foot of the Huachuca Mountains where Schieffelin had stopped for supplies, army scouts joked with him that he would not find any mineral wealth in that area; instead, "the only rock you will find out there will be your own tombstone" (or words to that effect).[6] In September 1877, when he filed his first Arizona mining claim, Schieffelin remembered that prediction and consequently named his claim "Tombstone."

On June 17, 1879, Schieffelin brought an old blue spring wagon to a halt in front of the Safford, Hudson & Company Bank in Tucson. Crowds gathered to watch the spectacle of what appeared to be an impoverished prospector delivering the first of dozens of loads of silver bullion from the newly discovered Tombstone mining district. His find assayed at just under twenty thousand dollars (nearly half a million dollars in today's currency). He was about to become a rich man.[7]

Schieffelin indeed had found his El Dorado in the Tombstone hills. With the assistance of his brother Al, partner Richard Gird, and a number of eastern investors, his find paid off big time. In March 1879, following a series of legal battles involving contested mining claims in the Tombstone region, the Schieffelin brothers sold their collective interests in the Tombstone Gold and Silver Mill and Mining Company for over a million dollars each.

---

5. Many popularized accounts exist. See Faulk, *Tombstone*, 25–46, and Underhill, "The Tombstone Discovery," 37–76.
6. For the quote, see Thrapp, "Dan O'Leary, Arizona Scout," 294.
7. *Arizona Daily Citizen* (Tucson, Ariz.), June 17, 1879.

At age thirty-five, the now-rich eligible bachelor then traveled to New York City, Chicago, Philadelphia, and even the nation's capital, Washington, D.C. Having been catapulted to celebrity status, he bought expensive, fancy clothes; stayed at the most cosmopolitan hotels; dined in the finest restaurants; and met many distinguished and interesting people, including President Rutherford B. Hayes. He sat for photographs and gave interviews to newspaper reporters and western-lore authors, who spun his tales into dime novels. But he remained unhappy and restless. Even with all his newfound wealth, Ed never could stay away from prospecting for long. He longed for mountains and deserts. Civilization had not dimmed his memories of the old days or his desire to return to the solitary life of a prospector. For him, prospecting was no mere occupation—it was his passion, his life.[8]

In 1882, after first contemplating a prospecting trip to Africa in search of gold, Schieffelin decided instead that he would mount an expedition into the largely uncharted territory of Alaska. Having carefully studied geologic maps, he concluded that there was a great "continental belt" of mineral wealth extending from South America up through Mexico and the United States, into British Columbia, and probably northward toward the Arctic. It was a region he simply had to explore. Schieffelin spent no less than twenty thousand dollars (a huge sum in those days) equipping a small party of hardy prospectors with only the finest provisions. He then chartered a boat that carried his companions from San Francisco to St. Michael, Alaska. On board was a little shallow-draft sternwheel steamer appropriately called the *New Racket*—the first independent steamboat in the region—which Schieffelin and his compatriots used to ply the waters of the mighty Yukon in search of rich prospects. After two long, hot, mosquito-infested summers and a single frigid winter when the Arctic cold dipped to 50 degrees below zero (−46°C), they returned to San Francisco empty handed. As he stepped off the boat that carried him and his band of men back home, Schieffelin summed up his experience when he told reporters that he had had enough. He "wanted no more of Alaska."

During a visit to Philadelphia in 1883, Schieffelin met the widow Mrs. Mary Elizabeth Brown, an actress in the Philadelphia Traveling Theatrical Company. They fell in love and shortly thereafter rendezvoused in La Junta, Colorado, where they were married. After a brief stay in Salt Lake City, Utah,

---

8. Turner, "Ed Schieffelin," 16.

they returned to California in the spring of 1884, at which time Schieffelin built his bride a fine mansion on Central Avenue in Alameda, across the bay from San Francisco. Later, they moved to Los Angeles and shared a house with Ed's brother Al. Ed and his wife briefly settled into a sedentary life, but all too soon he grew restless again. His true home remained the mountains and deserts of the West. There he longed to be.

The Prospector's Mindset

Despite their best hopes, prospecting was rarely a full-time occupation for Schieffelin and others of his disposition. Embarking on any prolonged prospecting adventure cost a lot of money; even the most basic outfit ran several hundred dollars. On one occasion in the early days, Schieffelin worked fourteen months before he had raised sufficient cash to buy two mules, saddles, guns, and the vital prospecting tools: picks, shovels, and pans. To raise the money for his many grubstakes, he worked as a freight and stage driver, chopped wood, made fence posts, and slaved in a mine. After outfitting himself and spending a few satisfying weeks or months in the gold fields, like so many others, he found himself penniless. Again, he would work for months at odd jobs to raise enough money to embark yet again on another prospecting expedition.

Whether striking out on his own as a youth with only $100 in his pocket borrowed from his father or mounting a full-scale $20,000 expedition to Alaska, the words "give up" were not in Schieffelin's vocabulary. While many others grew frustrated and abandoned prospecting for other more stable occupations, Schieffelin's attitude did not permit him to.[9] He always found himself drawn back to his first and only true love. Reflecting in his memoirs years after his lucky strike, he wrote, "Although I was economical, neither drank, nor smoked, nor gambled, nor spent money in unnecessary ways, I was no better off than I was prospecting and not half so well satisfied."[10] He had made up his mind at an early age that if he "ever accomplished anything it would have to be through the mines."

9. Faulk, *Tombstone*, 26.
10. An acquaintance once commented, "Schieffelin refrained from indulging in even light wines ... not because he was fastidious in his tastes or fanatical, but from a continuous lifetime of temperate habits." See Rice, "The Schieffelin Brothers Ed and Albert."

Success came to Schieffelin only after working "poor diggins" for many years. His secret, he confided in his memoirs, was that he "would follow no more excitements and pay no attention to anything that anybody else had found." Success, he concluded, came only when he decided to "try something of [his] own." In his memoirs, as well as in interviews with famed historian Herbert Howe Bancroft and numerous journalists after his Tombstone discovery, Schieffelin repeatedly warned others against "following the crowd," stressing instead the importance of "striking out on your own." Though it "may take a long time to get there," he boasted, "I finally did. Perseverance will generally win."[11]

Prospectors like Schieffelin carved their own particular brand of civilization into the mountain regions of Colorado and Nevada. They flocked to the Arizona desert and the inland empires of Oregon, Washington, and Idaho. Some traversed the wilds of Montana or the Black Hills of South Dakota, while a few plied their way up the freezing waters of the mighty Yukon River. Yet unlike the pioneer farmers who settled into rustic cabins in the hinterlands and cultivated millions of acres of virgin prairie sod, and in sharp contrast to merchants who lay in comfortable beds at night in their frontier towns, the entire West became the domain of prospectors like Ed Schieffelin.

Schieffelin's memoir reflects this distinctive lifestyle and provides an insight into the mindset of a prospector. For example, his recollections demonstrate that prospecting was generally a lonely and dangerous occupation, but one which enabled an individual to formulate a unique perspective on life. Like many others, Schieffelin confronted hostile environments and endured the elements, from the blazing summer sun in the southwestern deserts to the freezing cold of icy Alaskan winters. His recollections demonstrate how, in seeking mineral wealth, he was willing to contend with life-threatening diseases and "critters"—everything from mosquitoes and centipedes to bears and panthers. His writings suggest that his fear of ambush or attack by renegade Indians and his suspicion of Mexicans and Mormons were not atypical. It

---

11. Bancroft, "Edward Schieffelin: The Discoverer of Tombstone," 6. Schieffelin repeatedly told reporters of his hard-learned lesson: "I followed the life for a great many years and didn't strike anything. The trouble was I always followed the crowd. . . . Finally, I struck out for myself down in Arizona." See "Edward Schieffelin," transcript of story in *Clifton Clarion*, November 24, 1886 (MS 711), in "Edward Schieffelin Papers, 1878–1942," Arizona Historical Society. For other similar statements by Schieffelin, see transcript of story in *Denver Tribune*, undated, "Edward Schieffelin Papers, 1878–1942," Arizona Historical Society, and Rice, "The Schieffelin Brothers Ed and Albert."

probably represented commonly held notions and attitudes of Anglo-invaders who, alone in the wilderness, felt very much alone and isolated from the familiar landscape of civilization.

While prospectors generally led solitary lives, Schieffelin's story brings to the forefront other aspects of the prospector's mindset: comradeship and a propensity to stay informed on those "who done well." His tale reveals that something akin to a brotherhood of prospectors existed as each one kept tabs on the activities of their friends and their wide range of acquaintances, often over a period of many years. Though they did not band together annually in "rendezvous" as did some fur traders, trappers, and mountain men, the prospectors' brotherhood consisted of wanderers who periodically camped together, shared grub, and "swapped lies." On more than one occasion Schieffelin wrote glowingly of the companionship he enjoyed with other prospectors. Resting by the campfire under the flickering stars after a hard day's work, the prospectors found an opportunity to renew old acquaintances, make new friends, and to stay up late telling tall tales. These, he believed, not only helped make "time pass very pleasantly," but also contributed to the fullness of life.

Though prospectors welcomed periodic companionship, Schieffelin's autobiography also reveals a suspicious streak that manifested itself in the back of a prospector's mind. For example, unless a formal written partnership agreement had been drawn up, prospectors generally kept their mouths shut about their own potential "rich diggins" when in the company of others. As the campfire flames flickered, clever prospectors tried to weasel out information about the "prospects" of the region in an attempt to see if good "color" (gold or silver) was present. But experienced prospectors like Schieffelin generally knew to keep their own counsel, fearing that if there indeed were rich diggins and word got out prematurely, they risked creating a rush or "excitement" that could result in overlapping claims and possible loss of a fortune. Schieffelin never wanted to repeat his Tombstone ordeal, in which an argument over a title claim developed into a long-drawn-out legal battle involving the famous (and aptly named) "Contention Ledge Mine."[12]

---

12. Initially Schieffelin did not tell anyone, including his own brother Albert or his business partner Dick Gird, about the exact location of his first Tombstone find. He constantly reminded his partners of the necessity for secrecy until the claim was secured, as he was determined "to get to the mines first and have my choice of them." See Faulk, *Tombstone*, 41.

PORTRAIT OF A PROSPECTOR

Yet, in spite of ever-present underlying suspicions of another prospector's possible design on one's own diggins, Schieffelin's autobiography reveals another pervasive aspect of the prospector's mindset—an adherence to a frontier code of ethics. While individual standards varied, if Schieffelin's experience was typical, prospectors as a group generally adhered to the code. For example, since there were few law-enforcement officers present to enforce mining claims, it was this unwritten agreement that forced each miner to respect another's "stakes." To move a pile of stones that marked the limits of one's claim was a sin for which there was little sympathy and virtually no forgiveness.

Prospectors were also willing to help the genuinely less fortunate and would not hesitate to share grub, tools, clothes, and in Schieffelin's case even money. Poverty was commonplace in the West, and nearly every prospector at one time or another had experienced pangs of hunger and had occasionally been forced to wrap his blistered feet with tattered rags when shoes or boots fell apart. As a result, more fortunate prospectors were often willing to assist other members of the brotherhood who found themselves temporarily, in Schieffelin's words, "down in their luck."

Finally, Schieffelin's writings reveal perhaps one last significant ingredient in the prospector's mindset—a fear of quitting prospecting prematurely. The decision to give up the search for mineral wealth forever and settle down as a farmer or rancher, or to take up any number of city occupations was a traumatic one. For example, in one of the more telling episodes he recorded, Schieffelin describes every prospector's worst fear: that by abandoning prospecting too soon (as was the case of Schieffelin's acquaintance, to raise pumpkins), a fortune could be lost. For Schieffelin the episode—aptly titled a "Freak of Fortune" in his memoir—illustrated that very fear. His one-time partner's decision to quit prospecting proved to be not just a lost opportunity but also an economic tragedy.

In summary, Schieffelin's story suggests that some frontiersmen were born prospectors; no matter how many different jobs they worked in order to raise enough money for the next grubstake, they ever remained prospectors. Schieffelin's own family proved the point: his father, Clinton, was bitten by the gold bug at an early age and remained infected with gold fever until he died, and several of his sons (not just Ed but Eff and Albert as well) lived and died as prospectors.

Along about 66. during my first experience of Prospecting in the placer mines of south western Oregon. In a small Gulch that empties into a small creek calld Birdseye Creek. not far from louze lives. I thought I had found rich diggings. and accordin -gly built me a cabin, dug a ditch nearly two miles in length; and made prepra tions generaly for a winters work. but all for nothing. for I did nt make a quar -ter of a dollar out of it;

One evening. after being all ready to go too work, but having no water, for the rains that winter was late in coming. I had been hunting and was coming home. when not far from the Cabin I saw a smoke, which at first sight I took too be the Cabin a fire. but which proved to be couple of men, from an other mining camp about twelve miles from there. who had came on the creek that day a Prospecting. And seeing that somebody lived in the Cabin, and would most likely be there at night. concluded to camp there. so as too interview him that evening. and learn what they could concerning the prospects on the creek for diggings. And when I g

The first page of the memoir Schieffelin left behind. *Author's collection.*

The Schieffelin Memoirs and Autobiography

In death as in life, Ed Schieffelin remains the quintessential prospector. Beneath the stone monument on a small knoll just north of the city of Tombstone, Arizona, where his body rests, legend tells that his remains are dressed in a prospector's red flannel shirt, and next to him are his pick, shovel, and canteen—those simple tools that earned him his fortune. Also buried with him is the secret of the location of his legendary Red Blanket Mine, one of the few legitimate lost mines of the old West. But what he left behind for posterity is a record of his life and times—the Schieffelin memoirs.

After Schieffelin's premature death from heart failure while prospecting near his boyhood home in Oregon, his family discovered among his personal belongings a thin green journal containing a firsthand personal account of his life and adventures. It encompassed sixteen handwritten autobiographical essays ranging in length from two to twenty-one pages, all written sometime before October 1885. In the early twentieth century, the contents were shared with various members of the Schieffelin extended family by Ed's wife, Mary Elizabeth Brown. His sister, Jane Elizabeth "Lizzie" Schieffelin, eventually came to possess the memoir.[13] At her death it passed into the hands of their daughter, Neta Guirado Green.[14]

Upon Mrs. Green's death in 1978, the original journal could not be found among her personal belongings (apparently it had been passed on to or returned to another member of the family for safe keeping). But in a bottom desk drawer beneath a pile of yellowed papers, the executor of Mrs. Green's estate, Mary Jane Green Craig (Guirado's step-daughter), found a photocopy of Schieffelin's memoirs. That copy was probably made in 1958 or 1959, when

---

13. Born September 2, 1851, Jane Elizabeth Schieffelin married Dr. Ralph C. Guirado in Los Angeles on December 29, 1884. The nephew of Pío Pico (the last governor of Mexican California), Guirado was born into an old Spanish California family and at age twenty-five established Los Angeles's first pharmacy. A real estate promoter, he was active in politics and served as a founding member of that city's first Chamber of Commerce. See Neta G. Green, "Scrapbook," 7, 20, 32.

14. Neta Guirado Green was the only child of Elizabeth Schieffelin and Dr. Ralph C. Guirado. In 1904, after her father's death, she lived with her mother and grandmother (Ed's mother, Jane Schieffelin) in the house Ed had purchased. Miss Guirado's "Uncle Ed" was a frequent visitor to the household. She married several times. Her last husband was Daniel Webster Green, a renowned Los Angeles newspaperman. Mrs. Dan W. Green died January 19, 1978. See Green, "Scrapbook," 20, and Schieffelin Family Register, author's collection.

INTRODUCTION

Mrs. Green hired (and shortly thereafter fired) writer George Brandt to write a biography of Schieffelin based on his writings and Mrs. Green's recollections of her uncle.[15] It is the photocopy version of Schieffelin's memoirs from which some of this book is derived.[16]

Writing a scholarly or authoritative biography of Ed Schieffelin is beyond the scope of this volume. Aided by thoughtful editing of his memoirs and juxtaposing them with other manuscript materials narrated by Schieffelin himself, I concluded his life and adventures are best told by him. To this end, Schieffelin's "autobiography" is an artificial construct comprising his memoir vignettes organized chronologically, interspersed with several oral history interviews conducted with him by historian Herbert Howe Bancroft in 1886 and 1887. One such interview, titled "Edward Schieffelin: The Discoverer of Tombstone," provides much of the unifying thread between the various episodes.

In the pages that follow, Bancroft's transcription of his interview with Schieffelin is published almost in its entirety. However, in breaking up the text and interspersing it with Schieffelin's own writings, I hope I've been able to create a smooth chronological narrative that readers have come to expect of an autobiography. For the more scholarly minded, the footnotes provide a clear delineation where the Bancroft interview or each vignette begins and where another picks up the story. One episode that Schieffelin included in the memoir was also recorded by Bancroft in an oral interview: "Ed Schieffelin's Trip to Alaska." These two recollections, when combined with several memoir vignettes, give the most complete first-hand story in print of Eff and Ed Schieffelin's unsuccessful foray into the Alaskan wilderness.[17]

Schieffelin's last recorded "memoir" episode dates to 1885; however, he lived over a decade longer. While little is known of the last few years of his life, in

---

15. On October 13, 1959, Daniel and Neta Guirado Green signed a contract with author George Brandt, who was hired to collaborate with them to write a book tentatively titled "Ed Schieffelin's Life History, by Neta Green as Related to George Brandt." Mr. Brandt was introduced to the Greens by Mary Jane Green Craig's husband, Chase Craig, who was involved in the publishing industry. Brandt produced an eighty-page manuscript that ultimately was deemed unacceptable to the Greens. They discharged Brandt and abandoned the project. See W. Chase Craig, interview by author, Westlake Village, Calif., April 22, 1993.

16. In 1996 Marilyn Butler, great-granddaughter of Mary Elizabeth Brown Schieffelin, compiled and lightly edited some of Schieffelin's "adventures" and published them in her *Destination Tombstone: Adventures of a Prospector.*

17. Schieffelin's trip to Alaska is nicely chronicled by National Park Service historian Chris Allan in "'On the Edge of Buried Millions,'" 21–39.

the final chapter of this book, "Ever a Prospector," I've provided a thumbnail sketch of what is known. It pieces together the facts relating to Schieffelin's death of a heart attack in a lonely cabin near his last prospect some twenty miles from Canyonville, Oregon, and it addresses the myth and reality of Ed Schieffelin's "lost mine"—the so-called Red Blanket Mine.

Legend records that whoever finds the remnants of Ed Schieffelin's missing red wool blanket, shreds of which perhaps are still buried in the coarse dirt in the hills near Canyonville, may well unlock the secret Schieffelin took to his grave—a rich prospect that in a letter to his mother he wrote would make Tombstone "look like salt" (that is, a salted or worthless mine). "This is GOLD!" he reportedly wrote. But the rest of us must remain satisfied with his only other legacy: his writings.

Ed Schieffelin tells his life story in an entertaining and readable late-Victorian style. Generally devoid of histrionics characteristic of many period memoirs, he displays a remarkably sharp and perceptive memory. With a flair for vivid description, he manages to capture the excitement and camaraderie of life in the western mining frontier in direct, unsentimental prose. Above all, his autobiography reflects the spirit of the age and the drama of his life and discoveries.

Thankfully, Schieffelin had a penchant for the English language and a knack for clear expression. Consequently, except for the sake of clarity in a few instances, his writings required little substantive editing. He did, however, have an aversion to periods, and only rarely bothered to indent paragraphs, so in an attempt to improve readability and insure narrative clarity for the modern reader, some commas in the original have been replaced with periods, or at times simply omitted. Liberty has similarly been taken to ignore page breaks in the original manuscript. Likewise, capitalizations and a few connecting articles (and, on rare occasions, words) have been inserted and bracketed to insure the flow of language. In its original form, Schieffelin's writing is nearly always decipherable, but it occasionally falls short of literary correctness. Nevertheless, no attempt has been made to correct faulty grammar. Paragraphs have been recast only where it seemed logical and necessary to insure clarity of expression. For those unfamiliar with the language of prospecting, a glossary of terms associated with prospecting mentioned by Schieffelin in his manuscripts provides guidance.

INTRODUCTION

The editor would like to note that the debate relating to terminology commonly used for indigenous peoples in North America (including those of mixed race) remains controversial, there being little agreement even among tribal peoples. In Schieffelin's time, however, the word "Indian" was commonly used, though today the term is often viewed as pejorative. In keeping with the usage reflected in Schieffelin's writings and oral interviews, his use of the term has been retained.

One final editorial note: Schieffelin was generally a good speller. With the exception of "canon" for the Spanish word "canyon" and "iff" for "if," the manuscript has few consistently misspelled words. For the most part, his original spelling has been left intact. However, misplaced or omitted letters in proper and place names have been corrected. Annotations, the reader will note, are used sparingly. Except in the Tombstone and Alaska episodes, where other versions of Schieffelin's adventures exist and consequently warrant greater explanation, footnotes are limited to those deemed essential to put people, place names, and key events mentioned by Schieffelin into historical context.

Schieffelin once told a reporter that his "life experience . . . would have made an interesting book."[18] Having written his memoirs down with such care and precision, and having given so many interviews to historians and reporters, I suspect he planned to write his autobiography at some point in his life. My sincere hope is that this book tells the story he would have written down had he lived long enough to do so.

---

18. Unknown reporter, "Finding His Tombstone," in Neta Green, "Scrapbook," 22.

→ 1 ←

Early Life, 1847–1866

I WAS BORN IN TIOGA COUNTY, Pennsylvania, in October, 1847; the family was a large one, and came originally from New York City, where father was born.[1] My mother was born in the North of Ireland, whence she and her brother, Joseph Walker, were brought to America when very young.[2] Her brother came to California in 1849 and remained in this state until 1853, when he went to Oregon. During the Fraser River excitement of '56 or '57 he went to that country and after several years nothing more was heard from him.[3]

---

1. Bancroft, "Edward Schieffelin: The Discoverer of Tombstone," 1–2.

Edward Lawrence Schieffelin's ancestors settled in the American colonies prior to independence from Great Britain. His great-grandfather, Jacob Schieffelin Sr. (1757–1835), served with distinction in the Loyalist army during the American Revolution. Because of his sympathies, he fled to Montreal but after the war returned to New York City, where in 1794 he founded a drug company with his brother Lawrence. Ed's grandfather, Jacob Schieffelin Jr. (1793–1880), abandoned New York City for California during the gold rush of 1849. Unsuccessful in his prospecting efforts, he returned to New York City briefly but soon turned his attention to real estate speculation in Pennsylvania.

Born February 16, 1823, in New York City, Ed Schieffelin's father, Clinton Emanuel Del Pela Schieffelin (1823–84), joined the cavalcade of early western prospectors. After trying his luck prospecting in Wisconsin and Illinois, he traveled to California. Eventually he migrated north and claimed land along the Rogue River in Oregon. He purchased a parcel of land near Jewett's Ferry, where he moved his family in 1857. He died in Los Angeles of an accidental gunshot wound on April 15, 1884. For family history, see "Guide to the Schieffelin Family Papers," Beinecke Rare Book and Manuscript Library, Yale University, New Haven, Conn., accessed March 26, 2014, http://hdl.handle.net/10079/fa/beinecke.schieff. See also "Death Claims Pioneer of Section," "Obituary," and "A Neat Tribute," all in Green, "Scrapbook," 4, 32, 22.

2. Born in Ireland on May 23, 1829, Jane L. Walker Schieffelin (1829–1916) was brought to the United States at age four. She died April 15, 1916, in Pasadena, California. Little is known of her brother except that he may have traveled to California with Clinton Schieffelin.

3. In the spring of 1858, following glowing reports of great mineral wealth, some twenty-five to thirty thousand people headed for the Fraser River region of British Columbia. Gold was present, but in significantly less quantity than first reported. See Paul, *Mining Frontiers of the Far West*, 38.

Placer mining was the simplest form of prospecting. Using only a pan, gold could be extracted from gravel along a river bank. Woodcut, ca. 1860.

Until I was 9 years old I lived with my parents, brothers, and sisters on the farm in Tioga County, Pennsylvania. In 1856 we removed to California, where my father had gone in 1852. We arrived in San Francisco in the fall of 1856, and remained there and in Crescent City [Oregon], until about February 1857. We then went to Jackson County, southwestern Oregon, experiencing a very severe trip across the mountains from Crescent, about 100 miles, on mules and horses. The snow was so deep on each side of the trail as to reach in places the height of a man's shoulder as he sat on horseback. The wind was fearfully cold, and we suffered a great deal on the way; my mother, sister, and brother and myself, as well as my father, who had come down to meet us at Crescent City, were in the party.

On arriving in Oregon, the first thing I did was to get a shovel and a milk pan, and to go down the bank of the river looking for gold; that was my first experience with mica.

Finding particles of mica through the sand, and thinking they were gold, I stopped right there and tried to accumulate it. I started in the morning soon after breakfast, and by the middle of the afternoon had gathered probably half a teaspoon of the mica. I then came into the house, and met my uncle, who was quite delighted with my work, and seemed considerably amused. He told me the substance was not gold, and I then learned what mica was.

We remained for a few years on the farm, during which time once in a while I would take a pick, shovel and pan, and go off into the hills in the neighborhood, washing for gold in the gulches. Sometimes when out hunting, I would break the quartz and look for gold in it. This went on until I was about 12 years old, when the Salmon River mining excitement occurred, and I was seized with it and wanted to go there.[4] The fever was so strong that I ran away from home and got about 60 miles away. One of the neighbors brought me back. I remained home until I was 17.

My First Prospecting Trip, 1864–1865[5]

I well remember my first prospecting trip. I guess I was about seventeen. It wasn't my first prospecting by any means, but the first time I packed a horse and went out into the mountains to be gone any length of time. And then it was only for a couple of weeks and only about ten miles from home, but at that time I thought it was a long ways. I got to wondering how those that went through the country first must have managed, and what they probably thought when they was hundreds of miles from anybody and Indians everywhere.

It didn't seem so far nor as much of a trip as it did the first night or two. The first night the old work horse that belonged to Father that I had pressed into service to pack my outfit got loose by a coyote cutting the rope. The mule

---

4. In 1860–61 prospectors flocked to the Salmon River region of Idaho in search of gold. See Paul, *Mining Frontiers of the Far West*, 138, and Beal and Wells, *History of Idaho*, 1:266–324.
5. Schieffelin, episode one (untitled), *Memoirs*.

went back home leaving me entirely alone. My first impulse was to go after him, but on thinking over the matter for a while, I concluded that I was where I wanted to be, and had no immediate use for him. Anyway, if I got him again the same thing would most likely occur. So that I decided to let him go until I wanted either to go home or somewhere else—then I'd go and get him, for I was sure that he had gone straight home which I afterwards found to be so.

I guess I had been there about a week, sinking holes in gulches and creeks, and gouging around one place and another, occasionally finding a little gold but nothing of any consequence. Then, one evening about dusk, I went along up the main creek on which I was camped to see if I couldn't kill a deer. But it was too late so I didn't go far but came back to camp as I went without any deer.

Just up on the side of the mountain, a couple of hundred yards or so was quite a thicket in the head of a small gulch that ran down close to where I was camped. I was just starting a fire to get supper having set my rifle, an old muzzle loader, against a big pine tree within a few feet of where I was building a fire. All at once the damnedest racket I ever heard started right up in the thicket spoken of. For a second or two I didn't realize what it was and went for my rifle. My hat fell off and whether I knocked it off in my haste to get my rifle or whether my hair rising on end pushed it off, to this day I don't know! But I do know I was very badly scared for a little while in fact, until the fight ended, which it did in a very few minutes. I heard them run but it was too dark to see. But I heard them run over the point away from where I was, which was a great relief to me.

For it wasn't long after the fight began before I was satisfied that it was a couple of grizzlies fighting. Although I had never heard bear fighting before, the noise sounded to me worse that if all the dogs in the country was fighting at the same time, together with an occasional clap of thunder thrown in. I was sure it was bear.

I knew there was plenty of them there, although I had not seen any. Still I saw their tracks every day and panther as well. I would hear them scream almost every night and would sometimes think sure it was a woman's voice so much so that once or twice I answered, but receiving no reply, knew it was a panther.

In the morning I went up to have a look at the battle ground, but didn't find no dead bear. I had read that usually where wild animals that got to fighting, one of them would get killed. It wasn't so in this case, nor was there blood, flesh, nor any bones, but only here and there a little tuft of fur. But the brush looked as if a tornado had struck it, and the ground was somewhat scratched where they had stuck their claws in. Otherwise no one would never know there had been a bear in the country but for the noise they make, especially where there is heavy timber at night which makes up for all else—at least for a boy, that being his first experience with bear will think so.

I expected to meet with some of them while I was there but didn't. I remained there some time after that, long enough to satisfy myself that there was no mines there, nor has there been any found yet.

When I was about seventeen there was a mining furor in Montana, and I wanted to go there, but both my father and mother opposed, and finally I consented to give that up and go into the placer mines in our neighborhood.[6]

I went five miles off and went to work in a mining claim where I worked one month. Then, with what money I had made, I bought an interest in a claim, which I worked 2 or 3 months; then I quit on account of lack of water.

I next went to driving a team on the freight road for wages and continued at this until fall. In the fall I again prospected, and thought I had found a good prospect. I prepared for the winter, and dug a ditch to put water on the claim. When the winter came, which was late in the season, I worked a week or so, and found that the claim was not of any value. I gave it up in disgust. The winter had been a pretty hard one, and the water had not come until late, so that with the time I had lost, it was now well on towards spring.

I went to work for wages in a mining claim some four or five miles from where I had been prospecting. I worked there only about a month when the water failed. Then I went to work again on the road until toward fall when I went prospecting again. I found another prospect. It was in rather heavy ground and required considerable work to get at it. I built a cabin, and got tools, provisions, etc., ready for another trial. After working awhile, however, I concluded it would not pay, but I have since thought that I made a mistake, and that more than likely there was gold there, but in the bedrock instead of in the gravel. If I had worked up the bedrock I think possibly I might have got something out of it. As it was, however, I did not, but left that place, and during the summer and fall I prospected considerably in different localities.

---

6. Bancroft, "Edward Schieffelin: The Discoverer of Tombstone," 2–3.

Encounter on Birdseye Creek, 1866[7]

Along about 1866, during my first experience of prospecting in the placer mines of southwestern Oregon, in a small gulch that empties into a small creek called Birdseye's Creek, not far from the Rogue River, I thought I had found rich diggins. Accordingly, I built me a cabin, dug a ditch nearly two miles in length and made preparations generally for a winter's work. But, it was all for nothing for I didn't make a quarter of a dollar out of it.

One evening after being all ready to go to work, but having no water for the rains that winter was late in coming, I had been hunting and was coming home when not far from the cabin I saw smoke. At first sight I took it to be the cabin a-fire but it proved to be a couple of men from another mining camp about twelve miles away. They had come on the creek that day a-prospecting. Seeing that somebody lived in the cabin and would most likely be there at night, they concluded to camp there so as to interview me that evening and learn what they could concerning the prospects on the creek for diggings. When I got there I took them into the cabin, although they could have went in themselves for there was no lock to the door. As they had some grub with them I didn't give them any supper, but I did make their breakfasts.

During the evening we told stories, swapped lies and made time pass very pleasantly. I also learned that they had got there early in the day, about the middle of the forenoon and had been up and down the creek occasionally washing a pan of dirt. But they found nothing encouraging. One of them, the one who owned what tools (a pan) that they had brought with them, said that in the morning he was going back home as he didn't like the creek and didn't believe there was anything of any account on it. The other one seemed to want to stay a few days as I had told them that they could stay there with me as I had plenty of provisions. Since they had their own blankets there was nothing to hinder them from staying as long as they liked. But Shoe Butcher, which I afterwards learned was his name (getting it from mending boots), wouldn't stay.

The next morning, bright and early, he struck out for home. So I told Palmer, the other fellow (an Englishman), that if he wanted to try the creek farther that I would loan him a pick, shovel and pan and that he could stay

---

7. Schieffelin, episode two (untitled), *Memoirs*.

there in the cabin with me. There was plenty of grub, all he would have to do was cook it when I was away, which would be during the day for a few days but that I would be there nights. (I was going to help George Burns finish a ditch that he was digging. Burns lived on the left hand fork of the creek, some three miles above.) For all of this he seemed to be very thankful.

So I got the pick, pan, and shovel for him, took my rifle, and started out for Burns. When he shouldered his tools Palmer said that he would go along up with me and see Burns' diggings. He would then prospect from there down to the forks and up the right hand fork as he had been on it the day before and saw a place that looked well.

When we got to Burns and after looking at the claim, he went up on the ditch with us. After standing around awhile he went away. When he had gone George says to me, "That fellow won't do any prospecting."

"Why?" I says.

"Oh, he's no Prospector. He ain't shaped right."

That night or evening, when going down the trail, I thought I could see his tracks all the way down but thought nothing of it particularly; only it was a queer way of prospecting—to follow a trail all the time. But when I got to the cabin, I found out that he had done no prospecting for there set the tools up by the side of the door, perfectly clean, as I had given them to him in the morning. He had went in, cooked his dinner and I suppose ate it. It was gone at all events, and so was he. But the interior of the cabin looked like the devil, for he had been through and turned everything topsy-turvy, looking for something. What it was I don't know, unless it was money. I looked all around but could miss nothing whatever. Everything was there, only scattered around.

Then, when it was, I suppose, a couple of months afterwards that my brother Al came up one day and was looking at some quartz that I had brought in from time to time.[8] (When I was out hunting or traveling around and would see a nice piece of quartz, I'd bring it home and throw it down by the side of the door. I had accumulated quite a little pile.) Al thought he saw some fine

---

8. Born August 27, 1849, Albert (Al) Eugene Schieffelin (1849–85) was one of Ed's younger brothers and, like him, a prospector at heart. According to an acquaintance who knew both Schieffelin boys, "Albert Schieffelin possessed many of his brother's best qualities, he was amiable, unostatious [sic], of fine physique and withall an agreeable and accomplished gentleman." He died of consumption in Los Angeles at Ed's brother's home on October 15, 1885. See Rice, "The Schieffelin Brothers Ed and Albert," and Underhill, "The Tombstone Discovery," 51n43.

gold in one of the pieces and went to get a very fine magnifying or quartz glass that had been given me a year or two previous. I had always kept it in a little box nailed up to one of the logs over the head of my bed which served both as a shelf to set a candle on at night when I wanted to lay and read, and also as a shelf for small things. But the glass was gone. Al called to me that the glass was gone or that I had moved it. But I hadn't for I had not used it. We both hunted high and low and looked everywhere. We hated to give up the search because it was a very powerful and fine glass, one that I not only valued for its quality, but as a gift. At last it dawned on me that that fellow, Palmer, had taken it. Not being able to find anything else that was handy to carry and would be of any use to him and seeing it was an extra fine glass, he had taken it, feeling sure it would be some time before I missed it, unless by accident.

It was a mean trick. As Burns had said, he wasn't shaped right for a prospector. But, he was gone and so was the glass. I never saw or heard of him afterwards.

→ 2 ←

"Grubstakes"
Early Wanderings, 1866–1872

THE WINTER FOLLOWING [CIRCA 1866–67] was a dry one, and there was not much being done in the mines.[1] The next spring I went to Nevada, following a band of cattle in order to get there as far as Surprise Valley. From there on I went through the country with another companion, and we struck the Humbolt River, not far from Winnemucca. There I began my first prospecting for silver ores, and spent about two months there. Then I went to Owens River, California, prospecting on the way. At Owens River I went to work for wages, for about a month, chopping wood.

The Salt Lake excitement next attracted myself and three others, making a party of four. We started from the Owens River on New Year's Day, 1871, going down through the Pioche country through southern Nevada. The four who started traveled together three or four days, when—after we got to Montezuma (a small mining camp which had just started and where there was a demand for miners)—two of our number stopped to go to work in the mines. Myself and the other one of the party went on as we had started. We did some prospecting, until we got near Pioche, when we did quite considerable.

We then went on to the Salt Lake district, prospecting there until March, 1871, but we found nothing. My funds ran out, and I sold my saddle mule, put my blankets on my back, and packed them in and out of Salt Lake City, up to Corinne, and on up the Montana stage road, until I got up on Snake River. Here I went to driving stage, and did this for three or four months, getting a little money together. Late in the fall, I then went to Boise. I did not like that country, and came back into Nevada, where I spent the winter.

---

1. The first section of this chapter is from Bancroft, "Edward Schieffelin: The Discoverer of Tombstone," 3–4.

Encounter at Mountain Meadows, 1872

EDITOR'S NOTE: In their own version of Manifest Destiny to establish a New Zion in the American West, Mormons focused attention on the settlement of the Colorado River region. But in this quest, they were not alone. Some emigrants sought to settle there, too, while others would merely pass through "Mormon country" on their westward journey.

In September 1857 a group of about 120 men, women, and children from Arkansas and Missouri camped at Mountain Meadows, Utah, and were killed by Indians and Mormons. In the annals of western and Mormon history, the tale of the Mountain Meadows massacre remains both legendary and controversial.[2]

Schieffelin's visit through Mountain Meadows in March 1872 took place less than a year after Mormon officials had terminated the church's systematic expansion into the Muddy area. Still, some Mormon families remained in the area, often providing foodstuffs for miners and westbound immigrants.

Schieffelin's account depicts what he characterized as "that dark and bloody deed," a noteworthy description that reflects the prevailing negative attitude of many towards the Mormon settlers who, throughout the 1860s and early 1870s, were systematically establishing "stations" for migrating "Saints" from the East to the Mormon territory.

In March 1872, after spending the winter in Cave Valley Nevada about twenty miles north of Pioche cutting pickets to make a grub stake for the summer's prospecting, I started for the Grand Canyon of the Colorado.[3] On my way down, about thirty miles below Pioche, I over took a young Mormon who was going to St. George, southern Utah. It being on my road, we traveled together the balance of the way.

A couple of days or so on our journey we passed the Mountain Meadows made famous by that dark and bloody deed perpetrated by the Mormons, I think in 1858.

---

2. Several books today vie for the distinction of being "authoritative" about the event, including Brooks's *The Mountain Meadows Massacre* and Bagley's *Blood of the Prophets*. For a perspective more sympathetic to the Mormon leadership, see Walker, Turley, and Leonard, *Massacre at Mountain Meadows*. For an account perhaps best suited to the general reader, see Denton, *American Massacre*.

3. Schieffelin, episode three (untitled), *Memoirs*.

My companion, being a Mormon born and raised amongst them, knew pretty well their ways, although at that time he had begun to realize their evil ways and was beginning to be an apostate. As we rode along through the little green meadow of the mountains, he told me of the story of the massacre of the immigrants a few years previous and which was still fresh in the minds of the people of the world.

He began by pointing out the spot shown him as well as the story told by one of the participants who had long since repented and was sorry that he had ever had a hand in it. Where they camped was on a very pretty grassy slope about a half mile from where the battle occurred at the head of the canyon and where the doomed little band, entrenched by breastworks thrown up under their wagons, fought their worse than heartless foe all day.

Late in the afternoon as the sun was sinking behind the distant mountains bordering the beautiful valley of Los Angeles where they was going to make their homes, the Mormons, finding that they was unable to dislodge them without great loss to themselves, concluded to try the means of false friendship and thereby gain possession of their arms. When they ceased firing all became still, like the calm before a storm. John D. Lee, the leader of that treacherous and wicked band—and the only one among the many participants that was ever tried, condemned, and executed for one of the darkest deeds that was ever recorded in the annals of religious history—promised those poor half-famished creatures who were far from friends and surrounded by enemies of the worst type, that if they would give up their arms that he would take them safely back to Cedar City, about half way back to Salt Lake.[4]

The immigrants, after consulting, being out of water and surrounded by armed what they thought hostile Indians—some of their number layin' in the cold embrace of death, others wounded—concluded to accept those kindly offers by supposed friends, not knowing or thinking for a moment that it was Mormons disguised as Indians as well as Indians themselves that they had been fighting all day and which sent them all but two little girls to their graves.

As soon as the Mormons got possession of their arms, they marched them, men, women, children, one and all along before them up to where they had camped the night before. From a signal given by the leader, the slaughter

---

4. John Doyle Lee (born 1812) was executed March 23, 1877. No other suspects were indicted, brought to trial, or convicted. See Brooks, *John Doyle Lee*.

EARLY WANDERINGS, 1866–1872

A nineteenth-century illustration—one of the more famous images of the event—captures what Schieffelin characterized as the "acts of licentious barbarism" that took place at Utah's Mountain Meadows in 1857. *Courtesy Western History Collection, University of Oklahoma Libraries.*

commenced. Mormons and Indians vied with each other for the palm of cruelty sparing none but two little girls who afterwards were put to death in Salt Lake because they recognized one of their mother's dresses on one of the women in the family who they was living with.

The Mormons were ten-fold worse than the Indians in committing acts of licentious barbarism, unprecedented even in the imaginations of novelists. Not only did they tear infants from the breasts of weeping mothers, but by the heels dashed their brains out, the blood and brains bespattering the garments of the swooning mothers as they fell to the ground. Amidst that awful scene some were crying for mercy, with mothers shrieking for dead children clasped to their breasts. Above the groans of wounded, the death rattle of the dying, the religious curses of the Mormons, was the ever blood curdling yell amid suffering that was the whoop of the hostile Indians. And by the dead bodies of the fathers and under the staring gaze of the expired mothers, those incarnate fiends of human shape ravished young girls, some not yet in their teens. When not unconscious some would hold their struggling victim while others, as many as desired, accomplished their purpose. In the name of God they ended the suffering of the poor helpless creatures by cutting their throats then leaving the mutilated and exposed bodies for wild beast and disgusting vultures to prey upon. Then they gathered the livestock and turned it over to the church while the plunder they divided among themselves. My young friend said that at that time he knew a family that had one of the pianos and made their boast that it was taken from that train. The fate of the two little girls tells where the clothes of the party went.[5]

The Mormons claim that it was done for revenge, but it was done for plunder for it was the richest train that ever crossed the plains. The stock was all thoroughbred and at that time could readily be distinguished from the more inferior kind.[6]

It is hard to believe that such atrocious crimes could be committed by civilized human beings, by men who were fathers and husbands with daughters of their own, by brothers of innocent sisters and sons of loving mothers. But to those that has had experiences among the Mormons and knows the abject submission that ignorance and religion reduces mankind—their fanatic hatred for Gentiles or anybody antagonistic to any of their religious teachings, and the influence that the church has over them (that church ruled by a licentious avaricious revengeful man who is a vain absolute despot in whose breast, like the Kings of by-gone ages no spark of suffering humanity ever caused it one single pang of remorse)—can readily see that there is no crime too horrible for them to commit. Is it any wonder that crimes so revolting to more sensitive natures as to cause them to shed tears of pity for the victims and turn from the perpetrators with a loathing and disgust beyond description, are committed with an indifference and an abandonment of feeling that is perfectly indescribable.

---

5. Upwards of eighteen children survived the massacre and were temporarily placed in the care of local Mormon families. Schieffelin's reference to the fate of the two little girls may be a variation on a contemporary story that two teenage girls who ran to Lee for protection were sexually assaulted by Lee, then had their throats cut. See Brooks, *Mountain Meadow Massacre*, 101–5. For a detailed discussion of the fates of the survivors, see Bagley, *Blood of the Prophets*, 158–60, 236–47.

6. The wagon train consisted of sixteen wagons, one hundred oxen, and nine hundred cattle. Contemporary accounts considered it "the richest and best equipped train ever to set out across the plains"—perhaps a slight exaggeration but, nevertheless, Schieffelin's assessment seems accurate enough. See "Lee's Victims," *San Jose Pioneer*, April 21, 1877, and discussion in Bagley, *Blood of the Prophets*, 96, 102.

Whenever the dark mantle that enshrouds Utah is thrown off and she stands to the world in her true light, there will be revealed innumerable dark crimes and suffering that has been hoarse [sic] that will be beyond the power of tongue or pen to describe.

My Closest Call, Spring 1872[7]

In the spring of 1872 I had about as close a call as I ever had in the Virgin Canyon about sixty miles from St. George just below where the Beaver Dam Wash empties into the Virgin River.

As usual I was alone and the nearest white man to me was at St. George that I had left the day before. And if they had have known what I was I would have been glad had I met the fate that at first seemed there was no escape from. Afterwards it looked more miraculous to me than at the time.

Before I had got to St. George on the road a young Mormon overtook me and we traveled together three or four days. He warned me to do no prospecting in that country at that time. He said:

"You no doubt have heard (which I had) of parties coming down here into the Buckskin Mountains and never being heard of again. The Mormons could tell what had become of them if they was a mind to, so you go on about your business and nobody here would suspect you of being a prospector, but would take you for some cattle dealer from some of the mining camps up in Nevada down here looking for beef. If you keep your mouth shut they won't know the difference. There are lots of those Mountain Meadow massacre fellows in here one place and another. I could show you several of them as we go into and in St. George. I was born and raised a Mormon, but that don't make me one at heart. If they knew my mind in regard to them (although my parents are good Mormons) my chances would be slim. You folks think you know something of them, but you don't. So you take my advice, one who knows. Don't you do any prospecting nor talk about it while you are in this vicinity!"

Which I did to the letter.

No doubt you think by this time that my adventure was with the Mormons having said so much about them. But it wasn't. It was with the Indians. I mention the Mormons and what my companion told me so that you may know

---

7. Schieffelin, episode four (untitled), *Memoirs*.

what I mean when I say having traveled through the Mormons all right and getting where I thought I was beginning to be safe, met up with what I am now going to tell you in a very few words.

As I have already mentioned, it was in the Virgin Canyon close to Beaver Dam Wash. I had traveled all day without seeing any sign of Indians and would not have paid much attention to them even if I had, for those Indians had for a few years previous been considered peaceful, and I had heard of them committing no depredations.

It was a little before one of those beautiful sunsets that is seen no other place in the world but on those deserts in those southern counties but which I am not going to try to describe. The road all the afternoon had been running over a high mesa which virtually made the Virgin Canyon. Coming into the Beaver Dam Wash I followed it down into the canyon for some distance (the river as well as one side of the canyon is very crooked and thickly studded with willows so that an army of men might be in some of those nooks in the side of the canyon and on the opposite side of the river from where the road runs, and not be seen). So it proved with me for I had just entered the canyon and had crossed the river when through an opening in the willows I saw a lot of Indians who had apparently just got there and made camp.

Not having made any fire yet, nor was they on the look-out for anybody on such an unfrequented road as that one was (especially one man alone and probably the only one in six months before), we both saw one another at the same time. One Indian fetched a whoop which in that canyon sounded like thunder. Then, about sixty of them, all well armed with guns, started for me. Some got behind me and some ahead to keep me from making a dash for it. Long before I could have got where they filed across the road ahead of me even if I tried, they was there and coming to meet me at a slow trot. There was no use to undertake to go back for all they had to do was to climb over a point that divided the Wash from the Canyon and not over two hundred yards across to get to the road on that side, and I would have to ride in a half mile or so. So my only way was to put a bold front on and trust to strategy for there was too many to fight. Besides, I hadn't seen them all for I could still see some standing in camp and hear a great many holler.

They closed in around me, leaving a circle of about forty feet across for me to perambulate in. Considering the circumstances, my mustang was almost beside herself with fright—plunging, bucking and shyin' all at once keeping me busy to keep my seat. I believe that is what saved me for she drew their attention and I supposed amused them.

Before they knew what I was about, I yelled to them to "vamoose" (a Spanish word meaning "go") waved my hand for them to stand aside, and put my spurs deep into my mustang's flanks causing her to act like one mad animal. I made a break through the circle nearly riding one of them down before he could crowd himself into the crowd that was falling back to get out of our way. And away we went down that canyon like the wind, expecting any minute (until we was out of range and sight) to hear a gun go off and a bullet go whistling by, if not feel it. Nor did we stop until I was sure that I was out of all danger as far as they were concerned.

I rode until late in the night but not fast of course. I rode off the road and camped without unsaddling, with the end of the stake rope tied to my arm so that in case I fell asleep (which I didn't) my horse could not break loose from me, in case any of them undertook to follow me.

I started again before daybreak the next morning. I arrived at St. Thomas Nevada, where the Muddy and Virgin rivers come together, about two o'clock in the afternoon all safe and sound. The second day after my arrival, some parties on their way to Arizona (coming down the Meadow Valley Wash until it empties into the Muddy then down that to St. Thomas and on to Arizona) found where three men who had left Pioche a few days before them (all well armed and equipped with plenty of ammunition) had been killed by the Indians about forty or fifty miles from St. Thomas in the Meadow Valley Wash at a place called Indian Spring. The killings was supposed to have been done the third morning previous, making it the morning of the same on which I was surrounded in the afternoon. The two places were only about forty miles apart across the country the way Indians generally travel. Taking all the circumstances in consideration—that being the way they would naturally go to get to the Buckskin Mountains, their place of refuge—I am certain they are the same Indians that on that morning had killed the three men at Indian Spring in the Meadow Valley Wash.

Their action with me that afternoon was entirely different from friendly Indians. That was what scared me and made me run for it. If they had acted as they ought (had they meant no mischief) I should have stopped and got down out of the saddle and had a talk with them. But the moment I saw them move toward me, I knew something was wrong. Whatever kept them from firing on me before I had gone a hundred yards is more than I can tell.

↛ 3 ↚

Up the Colorado, 1872

AFTER GETTING A LITTLE MONEY TOGETHER late in the fall, I then went to Boise. I did not like that country, and came back into Nevada, where I spent the winter. In the spring I went to Arizona, stopping on the way at St. Thomas. From there I went up into the Grand Canyon of the Colorado, prospecting for placers, lost a boat, and had men drowned, and had to come back, making the trip a failure.[1]

Up the Grandest of All Canyons:
The Grand Canyon of the Colorado, March–April 1872

EDITOR'S NOTE: Interest in the mineral wealth of the Colorado River began with Spanish explorers. In 1827 trapper George Yount discovered gold in the river's lower canyons. He picked up a few gold nuggets, which gave rise to stories of the existence of a "lost Dutchman Mine."

Until the 1860s few miners had entered the region for fear of hostile Indians. Maps, usually deemed essential to help navigate treacherous waters, did not exist. In 1869, however, John Wesley Powell's legendary expedition into the Colorado River area (followed by George M. Wheeler's expedition in 1869–71) resulted in a comprehensive survey of the region. Not surprisingly, when Powell and Wheeler pushed up the river, they found trappers and miners already there. Nothing in the Schieffelin memoir suggests that he knew anything about the Powell or Wheeler survey parties when, in March 1872, he and a number of companions left St. Thomas on the Muddy River in the southeast corner of Nevada for their great—and eventually tragic—adventure up the Colorado River.[2]

1. From Bancroft, "Edward Schieffelin: The Discoverer of Tombstone," 4.
2. For early explorations of the Grand Canyon, see Smith, "The Colorado River," 112, 181–82, 338–43; Bartlett, *Great Surveys of the American West*; Powell, *The Exploration of the Colorado River and Its Canyons*; and Wheeler, *Preliminary Report Concerning Exploration and Survey*.

UP THE COLORADO, 1872

About the hardest trip I ever took was in the spring of 1872, when I, with four others, went up in the Grandest of all canyons, the Grand Canyon of the Colorado River.[3] Although it was a short trip—less than a month—it was a hard one.

It was along in the later part of March that there was about two hundred of us that left St. Thomas on the Muddy in the southeast corner of Nevada with two loaded boats on wagons all ready for launching, except caulking and pitching.[4] There was two parties with boats, four in ours and six in the other. The rest, when we got to the river we was going to ferry their outfits across, swim their horses, go across the country until they struck the Diamond River, follow it down and see if they couldn't get into the Grand Canyon that way. Our party, from some cause or other, was delayed one day at the Muddy which of course threw us behind in getting to the river. When we got there, some of the boys with the other boat was fishing and some setting around the camp fire. Only one man, Bush Dulin, was at work at the boat. The others with horses was a couple or so miles above us close to the mouth of the canyon. On account of the wagons us fellows with boats had to follow a big wash down which emptied into the river at that place, the only place where we in that neighborhood could get to the river.

At this time there was a party up in the canyon with a boat prospecting (and had been for some time) called Cook's party. Everybody was excited and speculating on what they had probably found. Some would gamble that they had found rich diggings or they wouldn't stay so long. And that if there was rich diggins anywheres in that canyon they would find it or keep going until they get to the Marble Canyon, something like two hundred miles up the Grand Canyon. When they got back to the river their feathers fell as they acted as if they

---

3. Schieffelin, episode five (untitled), *Memoirs*.
4. Established by Mormon pioneers in 1865, St. Thomas was a small village located near the confluence of the Muddy and Virgin Rivers; today it is under Lake Mead. The most likely route that Schieffelin and his party took began at St. Thomas, then they traveled southwest via Mud Wash, through St. Thomas gap, down Black Wash to Grand Wash Canyon (now Grand Wash Bay of Lake Mead). After reaching the Colorado River at the mouth of Grand Wash Canyon, he probably traveled up the Colorado to the general area of Pearce's Ferry. His party then would have entered the Lower Granite Gorge of the Colorado, at which time they paddled upstream before being turned back due to strong currents created by spring flooding. See David Miller, "Reader's Report: Portrait of a Prospector" for University of Oklahoma Press, February 25, 2016, in author's possession.

was sorry that they had started. They was setting around as above described as if waiting for something to turn up so as to have an excuse to turn back. There was nothing there to scare anybody only it was a dreary looking place.

It was too late to do anything that evening (the one we arrived at the river in), but the next morning the wagons went back and we went to work to cork and pitch our boat and make general preparations for the proposed trip. Yes, there was another boat with six men that I had forgotten about called the *Shamrock*. Mrs. Jennings, the wife of the man at St. Thomas who had furnished us with lumber to build our boats and such other supplies as she had when we had got our boats built christened them, calling the Irish boys' the *Shamrock*, ours the *Lady Jennings* and the other the *Anna*, her given name.

The *Shamrock* boys, all hands having gone to work the morning before was nearly ready to start. Their boat, that night before being in the water, was the first one launched, but they said that they would wait for us and we'd all start together.

That night we all had our boats in the river ready for loading. Dulin, by steady hard work, had succeeded (although having no one to help him—the others standing around whittling, talking of fishing) in getting his boat ready. He seemed as anxious as any one of us to go up the canyon, but all of his partners had long faces and had little to say to us fellows who seemed to be bent on going. Although there had not been a word said by anyone about turning back, still, it was plain enough that all but Dulin of the *Anna* wanted to and was only waiting for somebody to propose it when they would sanction it too quick.

The next morning, just as we was getting ready to load our boat and they was still sitting around the fire, an Indian came with a letter from Jennings at St. Thomas stating that the Cook party had got back after getting up in the canyon as far as the mouth of the Diamond River where they left their boat and such things that was too heavy for them to carry. They had gone into Arizona until they had struck the Old Beale Trail and had followed that to Beale Springs then to Mineral Park, the mouth of the Virgin River and thence home.[5] They had found nothing but had found it very hard and dangerous getting up the canyon and advised us all to come back.

5. In 1857–59 Lt. Edward F. Beale, best remembered for his "Camel Corps," which involved the experimental use of camels as pack animals in the Southwest, opened a wagon road that ran from New Mexico to California. The Old Beale Trail was heavily trafficked by merchants, prospectors, and immigrant travelers. Unfortunately, it lacked adequate water sources along the way, and parties traveling along it were also vulnerable to Indian attack. See Smith, "The Colorado River," 203–4, 220.

Well their faces brightened right up as if they had struck a rich prospect—just what they wanted. The Cook's party was all good prospectors, all on the work and had worked on the Snake River where it was all flour gold on the bars, just as it was there if there was any.

Before any of my party had a chance to speak, I turned loose. I says, "Don't let us turn back boys without giving it a trial. We have gone to a great deal of expense in outfitting and have got four months provisions. What can we do with it? There is no place to sell it. Besides, the boat will be a dead loss for we can't eat it nor do anything but leave it. Besides, there is a large party here, take us all, horsemen and all. And we don't know what they [Cook's party] might have found, nor but what they might have had some object in sending such word under the circumstances. Their grub must have been pretty short by this time, if not out altogether. They had to come in and get more and found good prospects but couldn't stay to work them. And on coming in and finding that there was such a crowd going, had sent word to discourage us to turn back. I am 'not' going back."

"Well," they said, "we will give it a trial" and to work we went to loading the boat. The *Shamrock* boys saw that we was going to go up the canyon so they loaded up and went.

While we was loading, Dulin came and sat down close to us. He looked awful blue, his face was as long as my arm. His party had all backed out so he couldn't go. He hove a long sigh, one that came clear up from his boots and said he was sorry that his party backed out for he was anxious to get up the river but didn't suppose there was any chance of goin' with us. I said that I had no objections if our boat had only been a little larger. He said his boat was larger and we could take that and leave ours behind if that was all which we all agreed to do. We loaded his boat, the *Anna*, and left the *Lady Jennings* on the sand bank.

With a little flag with "Good Hope" on it stuck to the bow, we took the letter the Indians had brought, and with the *Shamrock* boys along side of us, started up the river. When we got to where the horsemen was camped Oliver got up on a rock and read the letter to the crowd.[6] They all turned back to a man which left us and the *Shamrock* in possession of the field.

---

6. Oliver (most likely John "Jack" Oliver Boyer) was a fellow prospector who would also figure in Schieffelin's discoveries in the Tombstone district a few years later.

Just below the mouth of the canyon, about a half a mile, is a long riffle, not very rough but with a very strong current. In about the center of it, on the south side of the river (the one we was going on) was a large rock a few hundred feet in size, or, it was a portion of a cement hill that had become detached and had fallen into the river. The current being so strong made it a rather hard place to get around. Although we went by it all right, and we was ahead of the *Shamrock* we got over the riffle or rapids as it is most generally called, and was all still in the boat.

We was just going to enter the canyon (which is like going through a narrow gate in a very high wall) and as I was one of the rowers at that time and was looking down the river, I saw one of the *Shamrock* boys jump upon a rock and swing his hat and was acting in an excited manner generally. I says, "There must be something wrong with the *Shamrock* and we had better land, go down and see," which we did. The fellow with the hat was coming up towards us just a sailing and as soon as he got in hearing distance he hollered to Oliver who was ahead.

"Oliver, the *Shamrock* is sunk and all is going to hell. If yee's will set us across the river so that we can go back, we will be much obleedged to yee's."

We walked along down with him and he was giving the trip fits, blaming everybody connected with it in his Irish brogue which was somewhat amusing. For them boys it was very unfortunate.

When we got to the rock, sure enough, the *Shamrock*, in trying to get around that rock, had upset and everything that they had that would float was going dancing over the waves down the river. Blankets, clothes, and all such things was strung along down the river as far as the eye could see. They had saved nothing, only their lives and the boat. One poor fellow had taken off his clothes down to an undershirt and drawers, which was all he had. But we rigged him with a shirt and pair of pants.

As they still had their boat, they said that they didn't need us for as soon as they got it bailed out, they would all get into it and run down to where they started from that morning and the next day go to St. Thomas as it was only about forty miles from the river. They could easily make it in a day. So we left them.

Going back to our boat I looked back and there they were in the boat on the back track, leaving us alone, the five of us out of two hundred that had started out. From then on there seemed to be a kind of dampness to our feelings. We wasn't as jovial and lively as had been all the fore part of the day, but I don't think there was one of us that wanted to turn back or would have listened to such an idea had it been proposed. Still, there seemed to be a cloud over us.

When we got to the boat, instead of going into the canyon as we had calculated on doing before the misfortune of the *Shamrock*, we camped although it was still early in the evening. That evening we talked around the campfire which blazed up high and cheerful as if trying to dispel the gloom that had apparently settled on our hearts. But some approaching evil was shadowing our footsteps which plainly affected us all although none spoke of it. The only thing like an allusion to it was that Dulin who said, "Let us go slow but sure and above all, not get anyone drowned."

Little did he think, setting on that log of driftwood, the bright light of the fire shining on his honest face (as like the rest of us, trying to force a cheerfulness to dispel the foreboding spirit misfortune) that he was uttering a warning that ten days from that time was going to be fulfilled to the letter.

The next day, and for ten of them, we crawled up into that canyon under whose towering cliffs that for thousands of feet rear their tops toward the stars entirely shutting out the sun from casting its shadow on the river below, shutting us off as if it were from all civilization, leaving the world of life behind and entering into a realm in which animal life was still unknown. For the only living things we saw was two crows that followed us up the river and upon our leaving camp would fly down and pick up whatever they could find to eat.

Many times while going up the canyon on those rapids while working along with the boat trying to work it around rocks, my feet was swept from under me. But for good luck or something else I would always manage to grab hold of the boat or something that always seemed to be around, sometimes, looking as if it was there just for that purpose.

We would stop along and prospect wherever there was a sand bar that we thought was likely to have any gold on it.[7] Sometimes we got very good fine gold prospects, but the bars was very small and had very little dirt being mostly all rocks so that it took a great deal of ground as the fellows said to make a little dirt. I noticed the best prospects was in the edge of the water.

---

7. Because of their weight, particles of gold and iron often collect on sand bars along the banks and turns in rivers. Prospectors like Schieffelin often concentrated their prospecting efforts on or near these bars.

The lower we could get to the river, the better the prospect. But, it was the wrong time of year. The trip should have been made in the later summer instead of early April for the river was constantly rising from the melting snows in the mountains further up and towards the source of the river.

We got along very well until that day, the 10th of April. Close to the head of a heavy rapid and where the river was confined to a narrow channel, there was a large granite boulder, oblong in shape. I suppose it was twenty feet long, which lay with one end in the bar and the rest out in that boiling current making a very bad place to take a loaded boat. If we had unloaded the boat I don't think we would have had any difficulty. But we didn't and although there was four of us a pulling on the line when she took that current the force of it jerked us all in a pile and pulled the rope away from us. Down the river she went, leaving us there in that canyon in a hostile Indian country afoot. Nor could we see any way out. The walls rose up on all sides without an apparent break. When there was a break, there would be some wash that would run into the canyon which could sometimes be followed out—by that means a person could get out of the canyon. In most places, without a boat, that was the only means of getting out.

The boat didn't go far before she anchored by, I suppose the rope drawing in between the rocks. A large knot in the end prevented it from drawing through, holding it in the river not far from the opposite bank.

In the fall, Dulin hurt his knee. I was standing, talking to him when the other boys picked up the oars that lay on the bar and sung out to me as they started up the river to watch the boat. They would go up the river, build a raft out of driftwood, cross the river and then come down on the opposite side and get the boat. Dulin says to me that I had better go down and watch the boat as it might tear lose at any time. "As soon as my knee gets well," he says, "I will go up the river somewhere and swim over."

For a little while I tried to persuade him out of the notion for the water was very cold and there was a cold wind blowing up the canyon. But, he seemed confident of his ability to swim it, so I left him and taking my rifle went down on a point below the boat where I could see it up and down the river for a mile or so and the boat at the same time.

I hadn't been there but a few minutes before I saw George Magil come down on the opposite side of the river to where the boat was, but I thought it was Dulin. I could see by his actions that he could do nothing even though he was over there. At about the same time Charlie Goodnow came down on my side of the river and sat down on a rock opposite the boat. By this time I thought that the boat was there to stay so I would go and see what was going to be done and find out what had become of Dulin and Oliver.

When I got to where Charlie was, I saw then that it was George that had swam the river instead of Dulin. I wanted to know of Charlie what had become of the other two. He said that they was going to swim the river with the oars. At that same time I saw Oliver come running down the river as if the devil was after him. He said that Dulin was drowned.

On the opposite side of the river, from a little above the boat, up the river about a mile and a half above the rapids was a cliff three or four hundred feet high—a wall from the waters edge to the top. There was a shelf thirty or forty feet wide that once on it, there was no trouble in getting along. About half a mile above the rapids was a landing, a break in the cliff where it was easy to climb up to the top. At the break was a sand bar twenty or thirty feet long. But if that landing was missed by anyone trying to swim the river, they never would get out in the world because before they could anywhere near get back, the suction from the rapids would take them over. George, when he swam it, went high up the river and took a good start on it and had plenty of room and made it all right, but he said it was all that he could do.

Dulin and Oliver undertook it with the oars but didn't go far enough up the river and found before they had gone far that they was drifting too fast to make the landing so they came back. Dulin tried it again without any oar or waiting to get warm again and didn't go high enough that time either. Oliver said before Dulin had got three fourths of the distance across, when he drifted below the landing and was under that cliff when he sunk. Although nearly across there was no possible chance for him to get out or save himself.

George was over there but could do nothing. Charlie managed to throw his butcher knife over to him. Then from a pile of driftwood a little below, he got a long pole, tied the butcher knife to it and cut the rope close to where it was tied to the boat. It ran down a short distance to a short turn in the river where there was an eddy. He swam in and got it, took out one of the seats and with a hatchet made a paddle and came back to the side where we was on.

Our rope was gone, and without a rope it was impossible to get our boat up over the rapids, so we had to come away and leave Dulin in the river. We could go up on the top of the cliff and look down at the place where Oliver thought Dulin sank, but we could see nothing. The river is always muddy and very much so in the spring. There seemed to be an eddy where he sank, so no doubt he is there yet for they say the Colorado River never gives up its dead. It seemed so hard to have come away and leave him there in the river without trying to get him out. But, what could we do? We couldn't get our boat up there and without a boat it was impossible to get to where he had sunk. And to lose a comrade in that out of the way, wild, terrible and dismal place was horrible to think about. I imagined I could hear him calling not to leave him. Friends may die surrounded with comforts, or brothers or sisters where they can see the flowers and trees or at least fields of level country—even that causes us a great deal of grief—but to die in that dark and life forsaken place, and be left there was more appalling I believe than it is possible for it to occur in any other way or place, but it had to be done.

The next morning we got into our boat and ran part of the way down. A strong head-wind rose so that we was afraid to run for fear we might, in some of the rapids where the waves run pretty high, get upset, so we had to lay up until the next day. And then, before we got down to where we started we came near to being wrecked.

We was running the rapids with George on the lookout, I and Oliver rowing and Charlie steering when George yelled out "to the right to the right, a rock!" His warning was just in time, for the boat raked it as it went flying past. One inch more and it would have at least stove a hole in her if not mashed or upset us. We passed it all right, but I tell you, it was a close shave.

We got down about noon that day and that night went to St. Thomas getting in the morning for breakfast. I reported Dulin's death by a sworn statement before a notary who happened to be there at the time. He said he would have it published in a Pioche paper as soon as he got there. Then we rented a wagon out and brought in the other two boys and that ended the trip.

Lost a man and never made a single quarter.

I left St. Thomas, where we were had outfitted without a single quarter, and went down into Arizona, to Mineral Park with a man named Ike Ewing who paid my ferriage across the Colorado River which was 25 cents. There was not much being done except prospecting. I remained with Ewing and two or three partners of his for several months prospecting, but found nothing. I then got to work for some men who owned mines there and lived on the Colorado and worked six weeks. Then I went into what was called the Old Iritaba district, prospecting, and found some copper ores, but they were in small veins. From there, in the fall, I went to Prescott. The Indians were very troublesome all through that country; I did not find anything, and in fact did not think there was much chance to do anything in case I find a prospect.[8]

---

8. The last paragraph in this episode is from Bancroft, "Edward Schieffelin: The Discoverer of Tombstone," 4.

→ 4 ←

Southwestern Wanderings, 1872–1877

"Apache Territory," Summer 1872

EDITOR'S NOTE: As the trip down the Colorado demonstrated, prospectors on the western frontier not only experienced long periods of boredom but, by contrast, occasional moments of sheer terror. Threats, including renegade Indians and bandits, as well as environmental hazards like flash floods and poisonous insects, threatened their lives. Consequently, they had to maintain constant vigilance.

In a series of short vignettes, Schieffelin describes some of his various "close calls" in the battle between man and man, and man and nature.

About the hardest ride[1] I ever had was in the summer of 1872 in Arizona from old Fort Rock on the California road, by way of Hardyville,[2] to Camp Walapai.[3]

The Apaches those days was all over the Territory and it was dangerous to be safe anyplace. But the Prescott country was considered the worst of all (it was probably twenty-five miles farther on from Camp Walapai). Before getting to Camp Walapai the road passed over Juniper Mountain and down Juniper Canyon, a very bad place for Indians. Horsemen or parties of but a few men

---

1. Schieffelin, episode six (untitled), *Memoirs*.
2. In 1864 William H. Hardy established a landing seven miles above Fort Mojave. In the 1870s Hardyville hosted a population of about a hundred people. For miners in northwest Arizona, it served as a freight depot and way station for prospectors moving upriver or merchant traders going south. See Smith, "The Colorado River," 180–82, 224, 466–70.
3. Camp Walapai, a military camp, was located forty miles northeast of Prescott on Walnut Creek, below the Juniper range in Yavapai County. Originally established in 1869 and named Camp Toll Gate, it was renamed in 1870. Between 1869 and 1873 soldiers from the camp protected the roads to Prescott. See Brandes, *Frontier Military Posts of Arizona*, 40–45.

always traveled it in the night, it being much safer because Indians couldn't see you any distance.

At the time I speak of, the Walapai Indians, who for a short time had been on peaceable terms with the Whites, was in that section of the country hunting. Their fires at night could be seen in all directions. On going from Beale Springs to Fort Rock, I took chances and rode it in the daytime. During the day I saw lots of Walapais but paid no particular attention anymore than to occasionally stop and pass a few words with them as I usually done with friendly Indians.

When I got to Fort Rock, I found it only a station with four or five men but not a soldier in sight, nor had there been any there for some years. They all advised me to lay over there until some party should come along for they considered it at that time very dangerous. They was satisfied that those hunting camps of the Walapais was full every night with Apaches and it was very dangerous. But I had a wild mustang mare that could go as long as anybody's horse. Although I had never been in that country before, and the road was entirely strange to me, and a dark at night partially overcast with clouds, and fifty miles to Camp Walapai, I thought I could get through alright by taking the night for it and being cautious. For to wait there for a party to come along I might have to stay a month, at best a few days, and I didn't feel like laying there.

That evening, after it was good and dark (so that anybody watching the station from the top of any of the mountains in the neighborhood couldn't see whether anybody left or not), I started. My heart was about half way up in my throat but after two or three gulps I got it back to its place and rode along through junipers slow and easy I guess for an hour, not drawing a long breath for fear some of them might hear me. I spurred up into a lope and hadn't rode that way but a few minutes when suddenly I rode into a kind of hard pan. Any jump the mare made seemed to me like thunder; being fresh shod made it worse.

After riding some time with no improvement it begin to look that I was never going to get onto some softer formation. I had got so uneasy that the farther I went and the louder the sounds from her feet, the faster I went. When all at once, over my left shoulder (and not far off), I heard three keen whistles, like a man whistling through his fingers. And I had Juniper Canyon yet to go through!

I knew from what they told me at Fort Rock that there was some narrow places in it, places where a man on the side of the road could stick a butcher-knife in a fellow as he passed. For in some places the banks overhung the road and covered with thick brush. And I didn't know how the road run either. The way I had to go might be in a circle, while an Indian could cut across, get there ahead, and wait for me to come up. That whistle was a signal of some kind. There was no mistaking. And you bet I let out in earnest. If ever a man rode for his life I did. Nor did I slacken speed when I got to Juniper Canyon. How I got through that canyon without my horse falling has ever since been a mystery. I have been over the road since in daylight and it is not only steep but at that time the road was full of large boulders and holes washed out by recent rains. But we went down a-flyin', nor did the horse require any urging. I came out alright at an old fellows by the name of Brewster's, (about half a mile from Camp Walapai), at two o'clock in the morning having made fifty miles over a rough strange road in five hours on a Mustang that was worth about thirty dollars. And to all appearances she wasn't any the worse for the ride next morning.

Box Canyon Flash Flood, circa 1872[4]

If you want to get real badly scared just get caught in a box canyon once during one of these thunder storms or cloud-bursts as they are usually called in Nevada and Arizona.[5] Once will do you and you always afterwards will take good care to be out of them whenever there is one of those storms coming up.

To be riding down one of those box canyons so thick along the Colorado River—and in some places with no immediate place in which to get out as the walls on either side are hundreds of feet high, and hear the roar that you afterwards recollect—you can't mistake it coming along getting nearer and nearer. You have no idea how much farther you will have to go before there is a break in the walls so that you can get out and wait for it to pass, which usually only takes a few minutes. It don't seem to travel fast but if you undertake to get away from one once, you will think that it travels with the speed of lightning.

---

4. Schieffelin, episode seven (untitled), *Memoirs*.
5. Schieffelin is describing a flash flood, a sudden and often destructive rush of water down a narrow gully. In the southwestern desert, it typically occurs after a quick downpour of rain.

Often they say that under very dangerous circumstances all the mean things a man ever done in his life—in fact his whole life—comes up before him. But I don't think you would have to think much about your meanness when going down a canyon on the run with your pack mule with her head up and two or three jumps ahead of you and with every jump digging the spurs into the flanks of your horse trying to make him increase his jumps. As you look back over your shoulder and see that roll of water—mud, logs, sticks and rocks making a wall six or eight feet high, tumbling and rolling, sweeping all before it—and knowing that at least half a mile had to be made over rocks, cactus, brush, and such things without any road before you would have a chance to escape, you wouldn't think of anything but that desired place. Until you was out of the side of the gulch looking at that wall of water sweep by, would you know that you ever lived before.

I then went to work again for wages, driving a mule team, and thought I would quit prospecting.[6] I worked long enough to get sufficient money to take me back to Nevada. I started back to Eureka, Nevada, with the intention of doing no more prospecting, but to try something else. There was a good camp at Eureka, and a good deal of different kinds of business going on. For eighteen months I worked hard and tried several different things. Although I was economical, neither drank nor smoke, nor gambled, nor spent money in unnecessary ways, at the end of the eighteen months I found I was no better off than I was prospecting, and not half so well satisfied. So I made up my mind that if I ever accomplished anything it would have to be through the mines.

I left Eureka, and went to Austin, where my brother [Albert] who was afterward associated with me at Tombstone, joined me. We wintered together about 20 miles from Austin, chopping wood by the cord. My sole purpose was to make some money in order to go prospecting in Arizona. I partially accomplished this, when I became sick with a sort of mountain fever. I would not go to a physician, but moved around the country in different places. In the fall I finally found myself on the stage road from Winnemucca to Silver City, Idaho, leading horses for the stage company, my health no better than when

---

6. From Bancroft, "Ed Schieffelin: The Discoverer of Tombstone," 4–6.

I left Austin, and getting short of money again. I then decided that the trip home to Oregon, to that wet country, would cure me if anything would. In the latter part of November 1875, I left for home, arriving there finally, after going through a severe storm on the stage from Redding up, which left me entirely cured of the fever but with a bad cough. I arrived home with $2.50 in money, after being gone about six years in a country where wages was good and where men were making money; but I had not made any. I remained at home two or three weeks, then borrowed $100 of my father, and started out for Arizona.

The railroad then ran from Redding down as far south as Colton. I took the railroad and went on to San Bernardino, and as far towards Arizona as my money would carry me, which was to Ivanpah. I walked 60 miles on the stage road into Ivanpah with my blankets on my back, and $1.25 in money, and spent $1 for my dinner when I got there. I went there an entire stranger, to get another prospecting outfit together. This took me about fourteen months. I arrived there in December, 1875, and stayed there until about January, 1877.

It took that long before I had sufficient money to procure the kind of outfit I wanted, consisting of two mules, saddles, arms, etc., and all that was necessary to go into Arizona. I got these together with the determination of never parting with them again; I had a good outfit and made up my mind, no matter what occurred, I would keep it, and if I could not do any better I would live off my rifle until I found a prospect. I determined also that I would follow no more excitements and pay no attention to anything that anybody else had found, but I would try something of my own, and I did.

Escaped Lunatic, Summer 1877[7]

In the summer of '77 when [William] Griffith, [Albert] Smith and myself was doing the assessment work on the Brunckow Mine just previous to the discovery of Tombstone, we had quite an adventure—one I will never forget.[8]

---

7. Schieffelin, episode eight (untitled), *Memoirs*.
8. The Brunckow Mine (also known as the "Bronco" or "Broncho" Mine) was discovered by Frederick Brunckow of the Sonora Exploring & Mining Company in 1857. Just prior to the discoveries at Tombstone, Schieffelin met Albert "Alvah" Smith and William T. Griffith, who had been contracted by the owners to do the annual assessment on the unpatented Brunckow Mine. The assessment was a legal necessity in order insure that the owners could retain title to the mine. Schieffelin was hired to stand guard and assist the two assessors as needed. See Underhill, *The Silver Tombstone of Edward Schieffelin*, 20–21, and Frederick Brunckow File, Arizona Historical Society, Tucson, Arizona. See also Faulk, *Tombstone*, 10–21.

We were camped on the San Pedro, about half a mile from where the town of Charleston was afterwards built and where the trail from Cochise's stronghold crosses the river, running through a large thicket of willows.[9]

One evening, shortly after we had returned from the mine, I and Griffith was fishing while Smith was laying in camp, when happening to glance down the river, I saw smoke rising, evidently from a fire just kindled. Dropping my fishpole as if it had have been a hot poker, I ran to camp and there lay Smith about half a sleep with his gun sitting up against the small ash tree that grew there solitary and alone, serving us for a shade in the absence of a tent.

"What's that smoke just started down there in the willows on the trail?" I exclaimed.

"Don't know. It must be Indians."

"Well, let us go and see."

So taking his gun (mine always being in my hands and when eating, in my lap) we slipped cautiously along the bank through tall grass and low willows that grew along the sides. Keeping our eyes open and out of sight but always so that we could see the place in the willows where the smoke was slowly rising in a small spiral column. It was the very embodiment of a novelist's Indian fire.

Finally, we reached a place where we could see down under the willows, close to the water's edge. And there, in a spot clear of brush, about ten feet square, sat an Indian. Smith raised a gun to shoot him.

"Hold on Smith," I says, "hold on hold on. Don't be too fast. He hasn't seen us yet, nor can I see but the one. There may be a hundred of them in those willows. So wait a while and if there are anymore, some of them will certainly show themselves very soon. In the meantime we can watch this one's actions and determine what course to pursue. He can't get away from us. Besides, he might be a Mexican."

"A Mexican! The devil. What would a Mexican be doing there and naked? That fellow hasn't even got his blanket around him! He is one of old Cochise's damned Indians and I am going to shoot him."

"No you don't," I says. "No shooting to be done here yet awhile until I know more about him. Well, let's call him up to us."

---

9. For nearly a decade, renegade Apache Indians established strongholds in the Chiricahua and Dragoon Mountains. From their mountain camps the Apaches had a clear view of the desert below them. War parties ranged the country from San Siman to the Huachucas and from Gila to the Mexican border. See Burns, *Tombstone*, and Lockwood, *Arizona Characters*, 62.

"All right. We can step out so that he can see us and see that we are white men, and I'll cover him with my gun so that if he makes a break to get away I can kill him. If he should be a Mexican, he will only be glad to see us. And if an Indian and he sees that he is covered and can't get away, he may come to us. Once we get a hold on him, we are safe."

When we stepped out in full view, I drew my gun on him full cocked. Smith hailed him and at the same time motioned for him to come to us. Raising his head and seeing two white men standing out in plain view, armed with repeating rifles and one of them pointing directly at him, he must have thought the chances of running away was slim, and that he better come and see what we wanted. So rising up showing that he was perfectly naked and throwing his blanket around himself, he came indifferently and slowly along to where we were.

At the same time, Griff, who had heard Smith hollering had run to camp. Seeing the smoke rising from the willows, and seeing I and Smith down that way and an Indian crossing towards us, came running down to see what was up. He and the Indian both got there at the same time.

Turning to me Griff said, "Sheff, there are white men among these Indians. That fellow is no Indian, he is a white man."

Stepping up to him Griff asked him what he was doing in that country in the condition that he was. As from appearances he evidently was naked. If he had any arms none being visible but his blanket was wrapped around him and his hands underneath having the appearance of having a six-shooter in one of them and waiting an opportunity to use it. In answering Griff, he was impudent, telling him that it was none of his business whether he had any arms or not, or what he was doing in that country.

Then Griff told him that we was in a dangerous place, in bad Indian country, that his appearance created suspicion, and that we wanted him to camp with us that night. We would treat him well, but for our own protection under the circumstances not knowing who he was, we wanted him to raise his blanket and let us see whether he had any arms or not. And if he had, he must give them up until morning when we would return them as we got them. Reluctantly, he raised his blanket showing that he was without arms.

With the exception of a part of one leg of a pair of pants he was entirely naked, hatless, and his feet was wrapped-up in burley sacks. His skin was sunburnt and badly tanned from exposure until he looked as much like an Indian as it was possible for a man to and still not be one. His single badly worn blanket did the double duty of clothes and for a bed. An old frying pan without any handle and about ten pounds of flour completed his outfit.

Amongst us that evening, we fitted him out with clothes, minus hat and boots (not having any of those, only what we was wearing). But he looked more comfortable and much more presentable the next morning when he started for Camp Huachuca after thanking us very profusely.[10] There, he stopped for a few days then started across the mountains for Santa Cruz Mexico. That was the last I ever heard of him.

In talking with him, I found him a well educated man and that his name, so he said, was Timothy Malloy. That was all—he would give no further account of himself. His reason for being in such a destitute condition was that he was unable to obtain employment. The last place he worked was for John Sizemore in Jackson County, Oregon. And that after leaving there he had kept coming south until he had spent all of his money and was in the condition that we saw him. He had become disgusted with the Americans and was now going to Sonora to make his home with the Mexicans. He had heard that they was very hospitable and kind to strangers.

The conclusion that we came to was an escaped lunatic as he had no canteen and no means of carrying water. But he had sense enough to follow the rivers.

A year or so afterward, when I was down at the mouth of the San Pedro, I heard of him. When he came to a cabin, and if there was no one around, he would go in, eat all there was cooked. Sometimes there would be enough for three or four men, it made no difference. He either would eat it all or take it with him, but the general supposition was that he ate it. He would always take a few pounds of flour (but disturb nothing else) accounting for the flour that he had with him when we saw him.

---

10. Fort Huachuca was located on the northeast side of the Huachuca Mountains about eight miles south of Fort Walker and twelve miles from the Mexican border. It was one of the many posts established in the late 1870s to provide a measure of security to the region. See Brandes, *Frontier Military Posts of Arizona*, 40–45; Ferris, *Soldier and Brave*, 70–71. For an overview of the history of the fort, see Smith, *Fort Huachuca*.

Centipedes, Summer 1877[11]

I don't believe that a centipede's feet are poison, at least to have one crawl over you won't hurt you for I had it tried on me once. I don't know what might have happened had I hurt it so that it had have stuck its feet into my flesh as they say it does when either frightened or hurt. But even then I doubt it, for they have fangs and that is where I think their poison comes from.

It was at Tucson in the summer of 1877, and in that country centipedes grow pretty large for I have seen them that measured ten inches. I had been with some ore from near where the Tombstone district is now and I had been carrying it around town to see if I couldn't get somebody interested, but it was a no go. With very few exceptions they wouldn't look at it and those that did pronounced it very low grade. A man that would have put up $150 or $200 would have owned half of Tombstone, for with that much money then I could have found any mine there was in Tombstone of any account because there was no prospectors in the country and before anybody had found it out, I would have had them all.

Well, as I said, I had been poking around all day and had met with no encouragement. I had gone back to camp on a mesquite flat on the bank of the Santa Cruz River, a little way above Tucson. I had gone to bed and had taken most of my clothes off as it was very warm and being no danger of Indians. I was making myself as comfortable as possible under the circumstances and was laying there star gazing with only one blanket partially drawn over me. I was wondering what was the next best thing for me to do as I had found that Tucson was no place for a prospector. At least, not for me.

Suddenly, I felt something run up onto my left leg below the knee and start down for my feet which was naked, my drawers covering my legs and going as if it was in a hurry. Throwing the blanket off, just as it was going over my feet, by the light of the stars I saw it was a large centipede. And it felt large too! I raised out of the blanket like a flash of lightening thinking that it was two feet long and weighed a ton. And for a while I was scared and was on the point of going to a doctor, but kept waiting for it to hurt which it failed to do. It had passed over both of my feet and excepting the scare I got (which of course only lasted only a few minutes) I felt nothing of it nor did it show any rings on my legs next morning where it had been on them.

---

11. Schieffelin, episode nine (untitled), *Memoirs*.

Freaks of Fortune, Summer 1877[12]

Talk about the freaks of fortune and of chances that men have and miss, I will tell you of an instance that came under my observation, in fact, I was one of the parties.

When I was prospecting in the Tombstone country just before its discovery, one night, just about thirty miles from Tombstone on the San Pedro River where I was camped, W. J. Griffith and Alvah Smith came there on their way to the Old Brunckow Mine on which they was going to do the assessment work. It being a very bad place for Indians and as I had a good sharp rifle, a six-shooter, plenty of ammunition, and a field glass it was proposed to me that I would stand guard while they worked. I accepted.

The next day we went down the river where Charleston is—about a mile from the Brunckow where we camped on account of water (there being no water at the mine). After the work was done, Griff found that I was there in that country expressly for prospecting for I had not mentioned a ranch. Him and Smith both picked out places to take up ranches as they was coming down the river, and now as we was going back up the river, they was going to locate their ranches.

Griff proposed to me that if I would locate a claim for him (not as a partner), but an adjoining mine at the same time, that he would furnish me with provisions, have the recording done, and do what assaying that was necessary. He had a team which he could use to make enough money for that purpose.

I proposed to do it another way. If he would do as he had proposed, whenever I found anything I thought worth locating, that before doing so I would let him know it. Then one of us would build a monument at the places of discovery and the other take choice of claims. This was better and suited him better than the proposition he had made.

A day or two after that—after they had staked out their ranches of one hundred and sixty acres each—him and Smith went to Tucson to get provisions and money for their work on the mine, and some tools as that they wanted to build cabins and to work their ranches. They was gone about two weeks. I was there when they came back.

Griff says, "Sheff, I have heard you say often that fortune knocks at a man's door once in a lifetime. I think it is knocking at mine now." (At that time there

---

12. Schieffelin, episode eleven (untitled), *Memoirs*.

was a law called the Desert Act by which a man could locate six hundred acres of land, if he would reclaim it by putting water on it.) "There is a man in Tucson that says if I can find a good piece of land out here, to take it up under the Desert Act and he will furnish money for me to put water on it. And I am going to do it. It is a disappointment to you for our agreement will fall through for I can't do both, but it is a chance that I am not going to let pass."[13]

"Alright Griff," I said, "you know your own business best, but I think that you will miss it."

They paid me for standing guard and the next morning Griff started down the river with his team looking for a ranch. On a Mesquite bottom about four miles above where Benson now stands, on the San Pedro River, he took up his six hundred acres and stayed there about a month. But when the Indians got to killing people all around the country, he loaded up and went into Tucson, stayed there a few days, and then went up to the neighborhood of the McCracken Mines where he spent the winter.[14]

Then, in the spring, when the excitement broke out about the Tombstone mines, he went back to where Contention City is now and him and a man by the name of Bruno went to raising pumpkins for the very mines, that, had he carried out his proposition, he could have owned one-half interest in. But he had not only lost that opportunity but his desert ranch too.

About that time he was in Tucson one day when somebody wanted him to claim half interest with me in all my mines under that agreement. But he wouldn't do it. He said it was all his fault, that I had acted perfectly honorable with him and he should by me—that if he had have done as I wanted him to do, he would have had just as many and just as good claims as I had. But by his own act he had thrown the chance away and nobody but himself was to blame for it. So, when we sold out, I gave him five thousand dollars for his honesty. He then sold out and went to Virginia, to his old home, calculating to start a Tobacco store but I never heard from him.

---

13. Under the provisions of the Desert Land Act of 1877, settlers could claim up to 640 acres of land, provided they agreed to irrigate the land. A down payment of twenty-five cents per acre was required, with another dollar per acre due in three years. See Faulk, *Tombstone*, 212n3.

14. The McCracken Mine was located in Mohave County near the Bill Williams River in northwestern Arizona, some three hundred miles from the Globe and Silver King mines. In 1877 it employed about one hundred people. See Barnes, *Arizona Place Names*, 258; Underhill, *The Silver Tombstone of Edward Schieffelin*, 25; and Faulk, *Tombstone*, 35.

→ 5 ←

The Discovery of Tombstone, 1878

EDITOR'S NOTE: In 1877–78, after years of prospecting, Ed Schieffelin and his partners—brother Albert and assayer and mining engineer Richard Gird— staked out numerous claims in southeastern Arizona that ultimately netted them all fortunes. The Tombstone mining district became one of the most prosperous mining regions in the entire nation.

Ed Schieffelin's tale of the discovery of his claims at Tombstone has become one of the standards of western lore.[1] Most versions are based on contemporary accounts given by Ed Schieffelin, Richard Gird, and a handful of others to journalists and historians.[2]

Schieffelin titled the account that follows as "The History of the Discovery of Tombstone and How It was Named." It was probably written around 1883. Though it is far less detailed than the other existent accounts, nevertheless, it provides an insight into what Schieffelin thought was particularly important and chose to remember half a decade after making his famous discovery.

---

1. See, for example, Breakenridge, *Helldorado*; Burns, *Tombstone*; Faulk, *Tombstone*; Lockwood, *Arizona Characters*; Lockwood, *Pioneer Portraits*; and Underhill, *The Silver Tombstone of Edward Schieffelin*.

2. See Bancroft, "Edward Schieffelin, the Discoverer of Tombstone," and "History of the Discovery of Tombstone, Arizona, As Told By the Discoverer Edward Schieffelin," Arizona Historical Society, Tucson. Several published versions of the manuscript exist; one of the best is Schieffelin and Trawick, ed., *History of the Discovery of Tombstone, Arizona*.

For Richard Gird's account, see "Richard Gird File," Arizona Historical Society, and "True History of the Discovery of Tombstone," *Out West* 27 (July 1907): 35–50. An abridged version of Gird's may be found in Breakenridge's *Helldorado*, 89–98. Lonnie E. Underhill has also compiled the Bancroft, Schieffelin, and Richard Gird accounts into a useful single source; see his "The Tombstone Discovery," 37–76. Not all the accounts are in total agreement on some essential facts. See "Personal Recollections of John Vosburg to Frank Lockwood at Tombstone," an account in the Lockwood Collection of the Arizona Historical Society, which presents a slightly different account of the discovery. Of the diverging accounts, Faulk's *Tombstone* is perhaps the most accurate, as it relies on information substantiated by other sources when they diverge; see Faulk, *Tombstone*, 213n7.

In the summer 1878, I had really found the Tombstone district in Arizona, although I did not know then what I had found.[3] Fact is, I had found the ores of that district within six months after I had left San Bernardino with my outfit. After finding the district, I did not have any means either to have records made or assays, nor was there an assayer in that section of country. At Tucson, the nearest point, about 70 miles distant, there was none. I then started out and found my brother Al, in order to get somebody to assist me in making the locations. We then met with Richard Gird, who became our partner in the mines.[4]

We three returned in the spring of 1878 to Tombstone district, and located those mines. I had never before had a mine recorded, in all my wanderings, mining and prospecting. I had found a good many prospects, in different camps and excitements that I had followed, but nothing that I had any confidence in or thought I could make anything of importance of. Consequently, I never had a mine or claim recorded. Sometimes I would locate them, and build a monument, or put up a notice for the purpose of holding them for a short time until I could make further examination. But in every case I abandoned them. My long-continued ill-success was owing, I believe, to following other men's footsteps and following up excitements. If I had given that up sooner, I should probably have been more successful.

The History of the Discovery of Tombstone and How It was Named, 1877–1878[5]

In March 1877[6] I was prospecting in the Walapai Country along the line of where the Atlantic Pacific R.R. now runs, when a company of the Walapai Indians were enlisted to go to the southern part of the Territory on a scouting expedition against the Chiricahua Apaches who was committing depredations

---

3. The first section of this chapter is from Bancroft, "Edward Schieffelin: The Discoverer of Tombstone," 6–7.

4. Richard Gird was born near Cedar Lake in Herkimer County, New York, on March 29, 1836. He arrived in Arizona Territory in 1861 and is credited with bringing the first civil engineering and assaying outfit to the region. He knew the territory well, having prepared a topographical map in 1865. Gird was Ed and Albert Schieffelin's only non-family partner until

whenever the opportunity presented itself.⁷ Thinking that there was a good opportunity for prospecting, by going with them (for they would afford me protection). I went along but not as a scout as many supposes.

We arrived at Camp Huachuca about the first of April where the scouts remained for some time, recruiting and making preparations for the summer's campaign. During that time I would take trips through the country, notwithstanding the warnings I received from the soldiers of the danger of going alone. On my returns they would ask if I had found anything, which I had not. "You will," they would say, "You'll find your tombstone if you don't stop running through this country all alone as you are while the Indians are so bad."⁸ The remark being made often impressed the name on my mind, so much that the first mine of any importance that I found I called it "Tombstone," thinking at the time of the vast difference in the one I had found and the one referred to by the soldiers.

In May the scouts started out to be gone twenty days and I with them. But it wasn't long before I discovered that a scouting party was no place for a prospector. If I found anything, I must take the chances and go it alone for they was hunting one thing and I another. Consequently, our movements were entirely different. So on their return, when they got to the San Pedro

---

eastern capitalists invested in their claims and became partners. Like the Schieffelin brothers, Gird eventually sold his mining interests and moved to California, where he purchased the 37,000-acre Santa Ana del Chico ranch in San Bernardino County. There he ranched and farmed until his death at age seventy-four on May 30, 1910. See Faulk, *Tombstone*, 37–46, and Underhill, "The Tombstone Discovery," 37–76, as well as Bancroft, "Life of Richard Gird," 80–90.

5. Schieffelin, episode twelve, "The History of the Discovery of Tombstone and How It Was Named," *Memoirs*.

6. In other accounts, Schieffelin specifies the date "January 1877"; see Underhill, "The Tombstone Discovery," 44.

7. In 1872 the Chiricahua Apache concluded a peace with the United States government and moved to a reservation in southeastern Arizona. Early in 1877, however, Geronimo and a remnant band began what Schieffelin termed a series of "depredations" in southern Arizona. Though fewer in number than in the early 1870s, such Indian uprisings still gave prospectors cause for concern. See Underhill, *The Silver Tombstone of Edward Schieffelin*, 17, and Faulk, *Tombstone*, 23–25.

8. The comment is credited to several individuals, including well-known scout Al Sieber and a civilian scout, Don O'Leary. Most authorities now credit O'Leary with telling Schieffelin that, instead of silver, all he would find would be his "tombstone." See Thrapp, "Dan O'Leary, Arizona Scout," 294. For a discussion of the controversy over attribution of the remark, see Underhill, "The Tombstone Discovery," 45n14.

River one days march from the Post, they went on and I remained, not having found a color on the trip.

After I had been there a day or two, one morning before daybreak, a couple of men came along who had contracted to do the assessment work on the Brunckow Mine, located about eight miles from where Tombstone now is.[9] Seeing that I was well armed with plenty of ammunition they proposed for me to go with them and stand guard while they worked, the Brunckow Mine being a very dangerous place for Indians. Judging from the numerous graves around, together with the trail from Cochise's stronghold passing over the mine I thought their considerations correct and very suggestive of finding a tombstone.[10]

After finishing there we moved up the river a few miles where some others joined us.[11] All but me took up ranches which they thought very strange. Afterwards, when I had found rich ore and tried to explain and show them the metal in them, they would look upon me with pity and say they hoped it was so and that I had better give the prospecting up and take up a ranch that I would make more. And that if I found anything I couldn't do nothing with it anyways for it took capital to work a mine. I don't know what objections they didn't raise.

During the time that we were together, none of them would ever go with me to look at my prospects about twenty miles from there. I camped with them because I dared not camp in the Tombstone hills on account of its nearness to Cochise's stronghold. In fact, during the time that I was prospecting in there that summer, not once did I build a fire. I would ride in early in the morning taking a lunch and canteen of water with me and picket my mule near where I was at work. She was always on the alert, better than a dog, and seemed to have a realization of the danger she was in. I remained always saddled with the bridle hanging to the horn, my rifle in one hand and pick in the other, cartridge belt and six-shooter around me day or night, just the same. After

---

9. The two men were Albert "Alvah" Smith and William T. Griffith. See Underhill, *The Silver Tombstone of Edward Schieffelin*, 20–21.

10. "There was plenty of Indians. Old Cochise and his band were right around me... It was dangerous of course... all around in that vicinity white men had been killed that summer, but, being alone, I somehow escaped." See "Edward Schieffelin" in *Denver Tribune*, MS 711, Edward Schieffelin Papers, 1878–1942, Arizona Historical Society.

11. Most likely Schieffelin is referring to James Lee (1833–84), an owner/operator of a flour mill southwest of Tucson; the Bullard brothers (Charles M. and John C. Bullard); and George Woolfolk, all of whom camped out on the San Pedro River. These individuals are all mentioned by Schieffelin in his other accounts. See Underhill, "The Tombstone Discovery," 47–48.

THE DISCOVERY OF TOMBSTONE, 1878

doing this for a while, I would pack up and go off to some other part of the country for a week or so, return and try again, and so all summer.

In August, being out of provisions, no clothes and only five dollars in money (which in those days was like none at all), I took some ore and went to Tucson to see what I could do. I very soon found out for they would neither look at me or the ore and said that they didn't want any mines. Government contracts and Posts was good enough for them, and that a man was foolish to be spending his time looking for imaginary fortunes. He ought to be out at work somewhere—go take up a ranch and be somebody. That if a prospector found anything it would do him no good for it took capital to work mines, and before capital would invest, there had to be a great deal of work done to show what they were getting which all took money. One investment house (that not long after failed) gave me to understand that, unless I had money not to come there for anything, for if I did I wouldn't get it. I wanted a sack of flour damned bad, but as I hadn't the money I didn't go, nor haven't been yet. So I had to go back as I came, only a little mad and more determined.

A few miles from camp, I met a man named Sampson who was going prospecting for placer mines and wanted me to go with him.[12] As the Indians were on the warpath again, having killed sixteen not far from there a day or two before, all the boys were leaving the river, going to the different Posts, making the future look rather gloomy for me. I agreed to go. He furnished the grub. Next day we went to Camp Huachuca where I got mules shod having thirty cents left after paying for the shoeing. Then we struck out and was gone twenty-one or -two days and never found a color.

When we got back to the river, it was entirely deserted. Not a soul on it. Sampson, disgusted with the country, struck out for South America, but he didn't go far for I saw him a year or two afterwards. His leaving left me without grub again and he had none to divide for we had eaten it nearly all up on the trip.[13] But, I had a good rifle and plenty of ammunition and game being very plentiful—deer or antelope in sight constantly all day long—so I

---

12. Near present-day Fairbank, Schieffelin met W. H. Sampson, a member of the San Pedro militia. In another account, Schieffelin mentions meeting J. Landers as well. See Underhill, "The Tombstone Discovery," 48n32.

13. Sampson and Schieffelin prospected for no more than three weeks before parting company at the end of August. See Underhill, *The Silver Tombstone of Edward Schieffelin*, 22.

was all right. But there was no use for me to stay there any longer. I had found a good mining camp but had no money to have any records made so there was no use to make any locations. The truth of the matter is I had no paper to write the notices on, nor enough money to buy it with.

After studying the matter over ragged, with Indians committing depredations on all sides, and being hard up generally, I concluded the best thing for me to do was to see if I couldn't find my brother, Al, who was somewhere in the territory. The last I had heard of him he was at the Silver King Mine eight months before.[14] But where was he then? The Silver King was the place to go. If he wasn't there I could get on his trail and track him up.

So for the King I started—two hundred miles away through a hostile Indian country keeping in the mountains and away from roads and trails to avoid being ambushed. Then one day I came near meeting them while riding up a canyon where there was a spring of water. Not seeing any Indian signs until I got close to the spring where the first thing I saw was their camp. It was too late to turn back so I rode slowly on, ready to fire at the first one that showed himself. But they were all away, and being a war-party there was no squaws with them to give an alarm. By riding until late in the night, they couldn't track me up.

When I got to Globe City, thirty miles from King, I met a man that had been at work there. He said that they had discharged a lot of men, Al amongst them, and that he had gone to the McCracken Mine two hundred miles north.[15] And as my mules was sore footed, and having no money to get them shod, the best thing was to go to work. But where to get a job? There was but very little doing there and the town was full of idle men. The Stonewall Jackson

---

14. Schieffelin had not seen his brother Albert for four years but learned that he had been working in the Silver King Mine. Located in Arizona's Globe mining district (about five miles north of present-day Superior in Pinal County), the mine flourished from 1875 to 1888, when a drop in the price of silver forced its closure. In seeking out Albert, Schieffelin hoped his brother would have some money to have assays made of the ore samples that he had found in the Tombstone hills. See Underhill, *The Silver Tombstone of Edward Schieffelin*, 25, and Underhill, "The Tombstone Discovery," 50–52.

15. Schieffelin is commenting on the generally depressed state of mining in Arizona. In the spring of 1877, the first flurry of mining activity that had brought territorial status to Arizona in 1863 had passed, and the mines were in decline. While copper had been discovered near Globe, Arizona, in 1877 it was not being worked much. Several of the silver mines to the south and southeast of Tucson had also closed. See Faulk, *Tombstone*, 22. Discovered in 1874, the rich McCracken Mine was located in southern Mohave County. It had a short life, though, as by late 1879 the silver was exhausted and it closed. See Underhill, "The Tombstone Discovery," 52.

Mine, a few miles from there had just been sold to a California company and there were working quite a number of men.[16]

Thinking my chances were better, I went over there and there met some boys whom I was acquainted with. Some had been waiting for a chance to go to work for nearly a month and was then making adobes—making about enough to pay for their board by cooking it themselves and sleeping in their own tent waiting their turn for the promised job in the mine to come. Notwithstanding all that, I asked the foreman if he couldn't put me to work at something. But he said no, that he was refusing men every day looking for work.

Well, it was no use to stay there. Game was scarce and I hadn't had a square meal in so long that I was getting pretty slim.

The Best Blessed Meal, Fall 1877[17]

Little things make a fellow feel awful good sometimes. About the most pleased that I ever was in my life was an invitation to dinner once.

In the fall of '77 I was going from the Tombstone country to the Globe district. As the Indians was very bad, I had kept in the mountains so as to avoid trails and roads so that they couldn't lay and wait for me. And at the place I speak of is where I came down out of the mountains onto the San Pedro River and from there followed it down to the Gila.

I guess it was about ten o'clock in the morning when I got there. There was a woman herding cows out of the corn and doing her housework too while her husband, with some Mexicans, was husking and hauling the corn in to the granary. I unpacked, unsaddled, and staked my mule out and lay down under a mesquite tree not far from the house to smoke and rest. There was no use cooking for I had nothing to cook and worst of all I had no money to buy anything. I was in hopes that by my being close around when dinner came—in fact, I calculated to be talking to the man so that he would ask me to have dinner

---

16. The Stonewall Jackson Mine was located twenty-eight miles northeast of Globe. Discovered in March 1876, it prospered until 1885, when its silver was exhausted. In another account, Schieffelin refers to the Stonewall Jackson as the "McMullen Mine." See Underhill, "The Tombstone Discovery," 52.

17. Schieffelin, episode ten (untitled), *Memoirs*. This account served as the basis for a Columbia Broadcasting System (CBS) radio broadcast of the February 29, 1944, episode of "Death Valley Days."

with him. And there was a woman who would cook it so that it would be a real old fashioned country home dinner. The very thoughts of it made my mouth water.

So I watched the sun and that team in the corn field that they was loading. At last he jumped up on the wagon and in he come and drove up to the granary where I was as soon as he and commenced the conversation. I don't think I ever talked better in my life. It wasn't long till his wife came out and told him dinner was ready. But he kept on talking and, of course, I wouldn't give up my hold over him, especially since as he said nothing about grub. Standing there his wife seemingly got tired and turned around and went back to the house which was only a few steps and reminded him that the dinner was getting cold.

At last he started for the house without saying a word to me about dinner. I sauntered along towards camp with an extra long face and all that water in my mouth dried up. But I didn't blame him a bit for I was sure that had I told him that I was hungry or had suspicioned it, he would have asked me in.

He hadn't been in the house but a few minutes before he came out and called me in. I always thought and always shall think that his wife sent him back to ask me to come and have some dinner.

I went in and Moses what a dinner—a pot pie or stew, nice fresh bread, milk and butter. What a spread for a lean, lank, hungry man like me to set down to. I had been living on venison and who for two years had not sat down to such a table where there was plenty of fresh milk and butter. And didn't it taste good. Wasn't I the happiest man in Arizona! And didn't I eat. And she kept piling it on. I made no apologies for they wasn't necessary. She expected me to eat and I fully realized her expectations.

God bless her. I never saw her afterwards.

I then packed up and started again not knowing where to go to meet up with Al but took the trail back for Globe by way of the Champion Mine[18] three miles from the Stonewall.[19] I met the foreman coming down the hill going to the boarding house. I asked him for a job. I told him I had to have work. He surveyed me from head to foot a few times before he said

---

18. Opened in 1875, the Champion Mine was located twelve miles northeast of Globe. See Underhill, "The Tombstone Discovery," 52–53.

19. Schieffelin, episode twelve, "The History of the Discovery of Tombstone and How It Was Named," *Memoirs*.

anything. I guess I was a hard looking citizen for my clothes (what was left of them) was tied with strings and patched with pieces of blankets, gunny sacks or anything else I had been able to get ahold of and the patches had worn out.

"Well," he said, "judging from your looks you do need a job and that, very bad. I have no tools for to put you in the mine, but you can go to work on the windlass at nights.[20] Not a very good job nor very good pay—three dollars a night and you can get your meals at the boarding house, but you will have to furnish your own blankets which I suppose you have with you."

A poor job under any other circumstances, standing and turning a windlass all night, snowing most of the time, without shelter, for three dollars. But thirteen of them was enough—all the money I wanted. My mules was shod and rested, I had clean clothes on, grub enough to last me to the McCracken Mines and seventy five cents in money. What more did I want?

When I got to the McCracken Mines I found Al at work in the Signal Mine.[21] I showed him what ore I had with me. Like all the others that had seen it, he said it was of low grade. Well, I couldn't prove that it wasn't for I had not been able to have any assay made and only had my own judgement to go on and that went against everybody else's. Still, I wouldn't believe but what it was good ore, but we didn't argue the question. The foreman wanted a man to fill buckets, so I went to work that night.[22]

Not long after that Dick Gird came up to the mine as an assayer. Al, being acquainted with him, got him to assay my ore, it running from forty to two thousand dollars a ton and very easy ore to work.[23] Still, Al didn't say anything about going down with me. But Dick wanted to go but I wanted Al to go also. It was some time before I could get him in the notion, but at last he concluded to go.[24]

---

20. A windlass is a hoist used to lift the ore buckets out of a mine shaft.

21. Located on the west bank of the Big Sandy River in the Owens mining district, the Signal Mine was named for the Indians who sent up smoke signals. It was an extension of the McCracken claim. See Barnes, *Arizona Place Names*, 407, and Underhill, "The Tombstone Discovery," 53.

22. Albert got Ed a temporary job paying four dollars a day filling ore buckets at the mine. See Underhill, "The Tombstone Discovery," 53.

23. In another account Schieffelin calculated the assay at $40 to $600 per ton. See Underhill, "The Tombstone Discovery," 54.

24. After discussing the possibility of a rich discovery, Edward, Albert, and Gird formed a partnership in January 1878 and decided to go to the San Pedro country to complete an examination of the area. See Underhill, *The Silver Tombstone*, 29.

So we all threw up our jobs. Dick got a mule and light wagon, an assay outfit, and with my old guard, Beck,[25] harnessed with Dick's, we left the Signal the fourteenth of February about noon, and got to the Brunckow Mine on the twenty-sixth. There we camped for about three weeks, then built us a shanty out of the stalks of a species of cactus, using the wagon cover for a roof. No Indians being in Cochise's stronghold (old Cochise having died there a short time before and his Indians, thinking it bad medicine, had all left for other parts of the Territory), so that it was comparatively safe.[26]

When we left Signal there was another party that left a few days before us.[27] Nobody knew where they was going. No one but me knew where we was going and I wouldn't tell. At all the stations on the road to Tucson we could hear of them. It began to look as if they had found out where we was going. They all being acquainted with Dick and thinking that he had something pretty good. They was going to be there with him.

We reached the San Pedro River, and at the station, at the crossing and there they were. All but one was out in the hills, and he heard us go by. But it was a very cold, windy, disagreeable day, and, being sick he wouldn't get up to look out to see who it was, nor did we stop although we had calculated to before we saw their wagon at the station. Then we hurried by as soon as possible for we wanted nobody to follow us or be there until we had made our locations and those of the best.

Having to labor under so many disadvantages the summer before, I was satisfied that I had not found the best mines. After taking chances to hunt the country out and find the district, I didn't want somebody else to come in and get the richest mines. One of them told me afterwards that if he had ever seen us and Dick being along that he would have followed for sure.

While camped at the Brunckow and before we had been there too long (possibly a week—the mines that we had found up to that time wasn't of no very great importance) Dick says one day, "Suppose we load up and go over to the Chiricahuas where those other fellows must have went. They must have

---

25. Beck, Schieffelin's mule and longtime companion, died in Tombstone in 1881.

26. Cochise actually died considerably earlier—June 10, 1874. See Ewing, "New Light on Cochise," 57–58.

27. Schieffelin is referring to the White-Parsons party of three San Francisco prospectors: Josiah H. White, W. C. Parsons, and Albert "Alvah" Smith. According to Schieffelin, they "followed the same road as we were taking." See Underhill, "The Tombstone Discovery," 55–56.

had something pretty good to come all the way, not only from Signal, but from San Francisco down here to this country."

So, I saddled up old Beck and went into the Dragoon Mountains where Cochise's stronghold is and was gone a couple of days. When going back, not far from where the town of Tombstone is now, on the Westside Mine (but at that time was not [yet] discovered) I stopped and took a good look at the hills and the formations.[28] I made up my mind that if there wasn't good mines somewhere in that neighborhood, there was no use of judging from indications or experiences as they must amount to nothing. But if that wasn't a mineral country I had never seen one. If Dick was still in the notion of going to the Chiricahuas he could take his part of the outfit and go, but I would stay where I was. A day or two after that I found the Lucky Cuss.[29] Then there was no more talk of going away.

Hank Williams or Jack Oliver[30]—a couple of prospectors whose animals ran away from them and one of them came about fifteen miles from their camp and one got in with ours—and when they found it [the animal], seeing that we were prospectors with an outfit for assaying, thought that they would try it for a while as our assays showed that there was good ore there. After a couple of weeks prospecting and not finding anything they got discouraged and was going to leave when Dick persuaded them to try it a while longer. The result which was they found the Grand Central, one of the best mines in the district and one we could have had and would have had had Dick let them go when they wanted to.[31] As soon as they found it one of them went

---

28. The Westside was yet another of Schieffelin's claims in the Tombstone district.

29. Schieffelin, who had discovered a vein of almost pure silver about six inches wide and forty to fifty feet long, is making reference to the Owls Nest claim, one of five claims in the Lucky Cuss series. It was discovered March 1, 1878, with Henry F. Albert being credited with naming it the "Lucky Cuss." See Underhill, "The Tombstone Discovery," 57–59.

30. Born in England, Henry "Hank" D. Williams is the prospector credited with the discovery of the Grand Central Mine. He was also at the center of the controversy that embroiled the Schieffelin brothers surrounding the ownership of the Contention Mine. After the Arizona silver discoveries, John "Jack" Oliver Boyer murdered a man and escaped to Mexico. Upon his return to the United States, he was apprehended and sentenced to life imprisonment at the Yuma penitentiary. See Underhill, "The Tombstone Discovery," 59–61.

31. On March 27, Williams and Boyer located the Grand Central Mine. The next day Schieffelin located the Defense claim. According to Underhill, if Ed "had not stopped to help Albert and Gird move their camp, he would have located the Grand Central claim first." Underhill, *The Silver Tombstone*, 31. For a discussion of the controversy relating to Walker, Williams, and Gird, see Faulk, *Tombstone*, 47–51.

to Camp Huachuca, got drunk and told of the discovery which created an excitement.[32] The party that was in the Chiricahuas, White, Parsons and Albert Smith, hearing of it came in there. We sold the Contention Mine to White for ten thousand dollars.[33]

Not long afterwards we bonded our other mines to Charles Tozer for San Francisco parties for ninety-thousand, giving him ninety days to make the sale or lose the forfeit of five thousand dollars that he had put up.[34] The time given in the bond expiring and no sale, we went to work on the mines.[35]

We had been at work about two months when Governor Anson P. K. Safford came in and we gave him a fourth interest in our principal mines for putting up a ten stamp mill.[36] He, in order to raise the money, interested the Corbins

---

32. According to Faulk, Hank Walker showed ore samples to prospectors at Fort Huachuca, which in turn started the excitement. See Faulk, *Tombstone*, 50–51.

33. Located on March 29, 1878, the Contention Mine was aptly named. When Gird learned of Williams's and Boyer's discovery of the Grand Central, he found the two men had claimed the entire ledge and were not willing to share with the Schieffelin brothers as they had agreed. Gird knocked down all the corner monuments and claimed half of it for himself and the Schieffelins and named the claim "Contention Ledge." The Schieffelin partnership eventually sold Contention to Josiah H. White for ten thousand dollars. See Underhill, *The Silver Tombstone*, 34, 37.

34. A native New Yorker, Charles M. Tozer (1831–1905) arrived in Arizona in 1857. A mining superintendent by profession, he took a sixty-day option on the Tough Nut and Lucky Cuss mines but was unable to raise the ninety thousand dollars needed to consummate the deal by the agreed-upon option date of August 1. The Schieffelin-Gird partnership then turned to the Corbin family. Tozer eventually bought the Tombstone Mine (Schieffelin's first discovery) for twenty-five thousand dollars. See Underhill, *The Silver Tombstone*, 38.

35. There are several slight differences between Schieffelin's Arizona Historical Society account of the Tombstone discovery, this manuscript's versions, and the account left by John Vosburg, one of Schieffelin's business partners. There is nothing in either of Schieffelin's accounts to support Vosburg's claim that he was approached for grubstaking by Schieffelin and Gird before the three journeyed to Tombstone. For a discussion of this issue, see Faulk, *Tombstone*, 213n7.

36. Anson P. K. Safford, the former governor of Arizona Territory, arrived at Tucson on August 26, 1878. A native of Vermont, he knew how to prospect and had worked the gold fields of California and Nevada. Even though Schieffelin and his partners had sold some of their mines, they needed more cash in order to develop their remaining claims. In September, Safford and a Tucson gun merchant named John Selah, along with Daniel B. Gillette and John Vosburg, expressed an interest in investing in the development of Schieffelin's remaining mines. John Vosburg had come to Tucson in 1869 to establish an arms and ammunition business. He had served in the state House of Representatives and was territorial auditor for Governor Safford. Safford became Vosburg's "silent partner" and traveled back east to raise the money for the construction of a first-class stamp mill, a devise used to crush ore so that valuable minerals could be extracted. See Faulk, *Tombstone*, 53–55. For a detailed description of how the stamp mill operated, see *Tucson Citizen*, June 4, 1879; see also Underhill, *The Silver Tombstone of Edward Schieffelin*, 43–44.

In a desolate desert in Arizona, the silver-strike boom town of Tombstone came to epitomize the old West. *Courtesy Arizona Historical Society.*

of Connecticut, and by the first of June the following year (1879), the mill was running.[37]

Not being able to make arrangements with the stage company that by this time was running daily to carry our bullions on account of the risks and dangers from robbers, Mexicans and Indians, we had to take it to Tucson ourselves. As Al was in the East trying to sell stock,[38] and Dick Gird had to be at the mill all the time, somebody directly interested had to go with it

---

37. The Corbin family operated the Corbin House Hardware Company out of New Britain, Connecticut. Both Elbert and Philip Corbin were interested in investing in mines in Arizona. Philip Corbin returned with Safford to look over the Tombstone district. Eventually, the Corbins invested in a series of mining claims and held interests in the Tombstone Gold and Silver Mill and Mining Company, as well as the Corbin Mill and Mining Company. Philip Corbin's financing was instrumental in the development of the claims. See Underhill, *The Silver Tombstone of Edward Schieffelin,* 39–42.

38. On October 26, 1878, the Schieffelins, Gird, Safford, and Vosburg incorporated in order to raise more money to develop their mining properties. Albert Schieffelin went east to assist in the purchase of milling machinery and perhaps to arrange for the sale of stock for the Tombstone Gold and Silver Mill and Mining Company. See Underhill, *The Silver Tombstone of Edward Schieffelin,* 40–41.

so it fell to me. With four good mules, a thorough-brace wagon, driver, two guards on horseback (one a hundred yards or so ahead and the other the same distance behind) while I was on the wagon we were making two shipments a month without once being molested. This, notwithstanding the warnings we repeatedly got and of the danger, universally conceded by everyone in making large shipments at that time so close to the line of Mexico where in case of robbery they could get in a few hours' ride.[39]

In November, Al returned from the East. The Wells Fargo Company, having an office in Tombstone, ran messengers and shipped our bullion.[40] Then, I packed up and went out prospecting for about four months.[41] On my returning, I found a party from Philadelphia in Tombstone who wanted I and Al's interest [in the mines]. We let them have it for $600,000 to be paid in monthly installments, the last one to be on the first of September—we retaining possession of the stock until the last payment.

In 1878, after we made the location at Tombstone and sold one of the mines, the Contention, we made arrangements to have a mill built by giving a fourth interest in our remaining claims for the mill.[42]

After we put the mill up and started the work June 1st, 1879, we worked the mine and ran the mill until March, 1880.[43]

---

39. The first shipment of silver was transported June 17, 1879, by Schieffelin and Judge T. J. Bidwell to the Safford, Hudson & Company Bank in Tucson. The bank in turn sold it to the United States Mint. Each shipment Schieffelin delivered ranged in value from eighteen thousand to twenty-two thousand dollars. See *Tucson Citizen*, June 17, 1879, and Underhill, *The Silver Tombstone*, 44.

40. In November 1879 the Tucson and Tombstone Stage Company began hauling bullion to Tucson. See Underhill, *The Silver Tombstone of Edward Schieffelin*, 45.

41. According to Faulk, by the summer of 1879 Schieffelin was tiring of his role as businessman. In November he left Gird and brother Al to run the company while he prospected in New Mexico and Colorado. He returned to Tombstone in February 1880. See *Tucson Citizen*, January 23 and February 4, 1880; see also Faulk, *Tombstone*, 67.

42. From Bancroft, "Ed Schieffelin: The Discoverer of Tombstone," 7–8.

43. On March 2, 1880, Hamilton Disston, John L. Hill, W. H. Wright, W. E. Littleton, R. C. Tettermany, and H. G. Huey accompanied by W. A. Williams and Frank X. Cicott (the coiner of the mint in San Francisco) arrived to look over the Tombstone properties. Ed and Albert accepted $600,000 for their share in the Tombstone Gold and Silver Mill and Mining Company and the Corbin Mill and Mining Company. They also transferred title to several other claims. With the Schieffelin brothers out of the picture, the two companies were merged and a new corporation, the Tombstone Mill and Mining Company, was established. Once they sold their remaining two additional claims by the end of 1883, the Schieffelins had sold all their interests in the various Tombstone mines. See Underhill, *The Silver Tombstone of Edward Schieffelin*, 47–52.

THE DISCOVERY OF TOMBSTONE, 1878

In the spring of 1882 I started to go to Africa, and got as far as Philadelphia, when I received a dispatch from Richard Gird (who had been a partner with myself and my brother in the first working of the Tombstone mines) that he wanted me to meet him at Tombstone. He had a lawsuit then pending, and I was an important witness.[44] I therefore postponed my trip to Africa, and returned to Tombstone. I was there only a very short time, when I again contracted the malarial fever. I then decided instead of going to Africa to go to Alaska, as that was a cold, wet country, and since I had contracted the fever in a hot, dry country in Arizona, I therefore concluded that the change from one extreme to the other would cure me.

---

44. Schieffelin is referring to a lawsuit filed by the heirs of T. J. Bidwell against Richard Gird for $300,000. Bidwell and Gird had been equal partners at the time of Bidwell's death in July 1880. His heirs claimed half of the amount that Gird received once he sold his interests in the Tombstone mines—$600,000. Eventually, the suit was settled in Gird's favor, but the decision was appealed. Gird paid the heirs $100,000 to settle the suit. See Underhill, "The Tombstone Discovery," 42.

→ 6 ←

Alaskan Adventures, 1882–1883

EDITOR'S NOTE: In 1867 the United States acquired Alaska for about two cents an acre (around $7.2 million). In the words of Secretary of State William Seward, it was "a great storehouse for the nation." Prospectors had found gold in the Kenai River basin as early as 1850. Word of this and additional strikes along the Fraser River and elsewhere lured others into this northern mining frontier in the late 1850s and 1860s. Nevertheless, because there was virtually no published information about Alaska until 1885, only a few bold adventurers had dared venture there. But for a few prospectors like Ed Schieffelin, the Alaskan hinterland held great promise of discovery.[1]

After hearing of a discovery in the summer of 1881 of a rich prospect along the Tanana River, in March 1882 Ed Schieffelin, his younger brother Eff, and three others mounted one of the earliest and undoubtedly the best-financed prospecting expedition up the Yukon. Once again Ed followed his own best advice in "not following the crowd" when he mounted the $20,000 Alaskan expedition. Acting on the popular belief that Alaska possessed great mineral wealth, he concluded after studying geologic charts that a well-defined gold belt stretched along the Pacific seaboard from Cape Horn to the Bering Sea.

The trip to Alaska was a memorable one—enough so that he recorded three descriptive episodes in his memoirs relating to his adventures. From a historical perspective, Ed's chronicle of his sojourn is important for multiple reasons. First, the timing: one of the earliest accounts relating to the exploration of the Yukon, it was recorded years before the great Klondike rushes of 1896 and 1898, and it predates most other early prospector accounts. Second, collectively these three

---

1. See Wharton, *The Alaska Gold Rush*; Hunt, *Alaska*; Sherwood, *Alaska and Its History*; and Brooks, *Blazing Alaska's Trails*, especially chapter 18, "Notes on the First Prospectors on the Yukon," 311–34 (the Schieffelin trip is discussed briefly on pages 324–28). For a fine general overview of the history of the Yukon River, see Webb, *Yukon*.

Alaska vignettes provide a rare ethnographic insight into the customs and ways of Alaska's indigenous peoples as perceived by whites in the Victorian era.[2]

There are several individuals who were not part of Schieffelin's party but accompanied him on part of the trip. These travelling companions included Royal Berlin Museum ethnologist Johan Adrian Jacobsen and *New York Herald* correspondent and freelance writer H. D. Woolfe. Jacobsen's observations, including his colorful account of his voyage (only part of which was made with Schieffelin), was originally published in German and later translated into English.[3] In spite of his relatively short time with Schieffelin's party, Woolfe managed to publish two articles about his Alaska adventures.[4]

The transcription that follows is a composite of Schieffelin's various manuscript accounts of the expedition. Important supplementary or explanatory details drawn from the Jacobsen account and some additional details from the Bancroft Library version, which is generally less detailed, have been incorporated into the annotations and, in the case of the Bancroft manuscript, in rare instances into the text itself.

In the month of March, 1882, my brother E. L. [Effington] Schieffelin and myself, went to Juneau City, Alaska, with the intention of prospecting for the summer.[5] But when we reached Juneau we found that we were not near the section we desired to examine. In order to get there from Juneau we would have to cross the mountains. Then, after getting over on to the headwaters

---

2. Michael Gates, *Gold at Fortymile Creek: Early Days in the Yukon* (Vancouver: University of British Columbia Press, 1994), 18. The Bancroft Collection at the University of California–Berkeley contains a less-detailed version of the trip entitled "Ed Schieffelin's Trip to Alaska." See Bancroft, "Ed Schieffelin's Trip to Alaska, 1882–1883," manuscript, hereafter cited as Bancroft, "Alaska." For a published and lightly edited version of the ten-page Bancroft manuscript, see Drysdale, "From Tombstone to the Yukon," 10–15. For a more recent account of Schieffelin's Alaska journey, see Allan, "'On the Edge of Buried Millions,'" 21–39.

3. An English translation of Jacobsen's 1884 account titled *Alaskan Voyage* fills in many details of the Schieffelin expedition. For other brief accounts of this expedition, see Mercier, *Recollections of the Yukon*, and McQuesten, *Recollections*, 11–12. For an interesting contemporary account of people and events along the Yukon only a short time after Schieffelin's visit, see Schwatka, *Along Alaska's Great River*, especially 86–87. Schwatka missed Schieffelin on the river by days but was familiar with his adventures.

4. See "On the Yukon River," *Harpers Weekly*, November 3, 1883, 694–97, and "Among the Muhtes of Alaska," which appeared in *Frank Leslie's Popular Monthly* 17 (January–June 1884): 676–87.

5. From Bancroft, "Alaska," 1, and "Edward Schieffelin: The Discoverer of Tombstone," 8.

Schieffelin (*center*), along with his brother Eff (*second from the left*) and close friends Jack Young, Charles Sauerbrey, and Charles O. Farciot, posed for this picture with high hopes that their forthcoming adventure into the Alaskan wilderness would prove fruitful. When they finished, they came back satisfied that they wanted "no more of Alaska."
Courtesy Arizona Historical Society.

of the Yukon where there is a portage of probably 80 miles to make across a high range of mountains, we would have to get along as best we could with such clumsy boats as we could build out of whipsawed lumber. The country was also covered with snow, and I saw that I could not do anything there. For this reason we came to the conclusion that we would come back, and did so, returning on the same steamer by which we had come, to San Francisco.

My brother Eff went down to Tombstone and got three other boys while I, with the help of some others, built a small stern-wheel steamer, of about 15 tons burden.

We chartered a schooner and went to Alaska again, taking general supplies, provisions, tools, and everything necessary for a prospecting expedition, intending to stay three years. We left in the spring and arrived in Alaska in July, remaining there until the fall of 1883.

Alaska Prospects, 1882–1883[6]

In the Spring of '82, Old Charlie Farciot,[7] Jack Young,[8] young Charlie Sauerbrey,[9] my brother Eff[10] and myself left San Francisco on the schooner *H.L. Tiernan* that I chartered bound for Alaska on a prospecting trip with a little steam wheel river steamer on deck.[11] The reason we built the little steamer

---

6. Schieffelin, episode thirteen (untitled), *Memoirs*.

7. Swiss-born Charles O. Farciot was an engineer by profession but photographer by avocation. He originally met Schieffelin in Arizona and agreed to accompany him on the trip to Alaska. In addition to serving as engineer, Farciot compiled an album of pictures that documented the trip. Many of Farciot's photographs of the Alaskan adventure are in the Huntington Library collection titled "Schieffelin Brothers Yukon River Prospecting Trip, 1882–1883," accessed February 7, 2016, http://cdm1.1ibrary.uaf.edu/cdm/compoundobject/collection/cdmg21/id/961/rec/1. Farciot died of a heart attack October 27, 1891. See Rowe, "Arizona Views by Charles O. Farciot," 373–90; Mercier, *Recollections of the Yukon*, 34–35n10. For Farciot's photographic career, see Jeremy Rowe, "Following the Frontier From Arizona to Alaska: The Photographs of Charles O. Farciot," accessed April 5, 2014, http://vintagephoto.com/reference/ChFarciot%20Article.html.

8. Jack Young, like Schieffelin, was a lifetime adventurer. During the trip to Alaska, he wrote a series of letters that are now in the possession of the Bancroft Library. In one, he describes Alaska as "intirly unsivelised . . . only 12 white men in the country, and white girls there is none." After returning from Alaska, he went to Mexico to search for legendary lost treasure; while there, he caught yellow fever and died in 1885. See Young, "Jack Young's Adventure," 16.

9. Schieffelin spells the name "Sawbury," but according to most authorities, the name is Sauerbrey. Little is known of his life. At twenty-seven years old, he, like Schieffelin's brother Eff, was among the youngest of the expedition. He was recruited with the others by Eff in Tombstone. For a brief biography of Sauerbrey, see *An Illustrated History of Southern California*, 366–67. See also Bancroft, "Alaska," 1–2.

10. Effingham Lawrence Schieffelin (1857–1933), one of Ed's younger brothers, died in 1933 at age seventy-six in a cabin fire while working a claim. See "Miner 76 Who Died in Blaze Laid to Rest," in Green, "Scrapbook," 38.

11. Rather than take the inland passage through Chilkoot Pass to the Yukon, which had been discovered in 1879, Schieffelin hired the captain and crew of the *H. L Tiernan*, a 153-ton two-masted schooner built in San Francisco in 1867. The ship sailed to Alaska via a tedious 2,300-mile route over the Pacific Ocean and choppy Bering Sea to the windswept, treeless, and mossy shore of St. Michael on the Norton Sound. From there Schieffelin planned to challenge the stiff Yukon current by using his little steamer to laboriously push its way up the swiftly flowing river.

was because in that country all traveling has to be done either on foot or with canoes and to get around with canoes, by means of paddles, etc., is hard and slow work. There are a great many navigable streams there and a small, light-draft boat like that we thought we could travel almost anywhere with and could go quickly. There was plenty of wood in the country and we would never need to be short of fuel, all that was necessary being to cut it.[12]

We went from San Francisco to St. Michael, within 60 miles of the mouth of the river with this schooner and were 40 or 45 days getting there.[13]

Everything went lovely going out of the bay until we got out about forty miles when the breeze went down leaving us in a dead calm where we remained for three whole days. Then a light storm come up bringing plenty of wind so that we done very well the balance of the way.

At Unalaska we put in to clear the customs house where the Alaska Commercial Company's boys gave us a dance.[14] Of course the girls were all Russian half-breeds, or natives, but they was all good dancers. But what struck me most was to see those Indian Girls dressed up, some of them in silks and all in the latest fashions, with full backs, cross-hitches and ruffles, hair banged, with kid gloves on, skin white, giving every appearances of white society ladies.

But they could only speak the Russian language and we couldn't so there was not any flirting going on. Still the boys danced even if they couldn't talk. Young Charlie told me they was splendid waltzers. A person's impression going in there for the first time is that he is going into an ordinary village of a few hundred inhabitants of white people, for the natives has that appearance until

---

12. Schieffelin's paragraph-long explanation of "The reason we built . . ." is from Bancroft, "Alaska," 1–2.

13. According to Jacobsen, "Our departure from San Francisco took place at noon on 13 June 1882. There was a large crowd on the wharf who expressed their good wishes by cheering and waving handkerchiefs. We had decorated the ship and small steamer on its deck with flowers and other trimmings, and great hopes swelled the breasts of all as we were towed slowly and majestically toward the Golden gate. 'On to the Yukon!' was the cry." See Jacobsen, *Alaskan Voyage*, 82.

14. After thirty-three days at sea the *H. L. Tiernan* arrived at Unalaska, an Aleut settlement that during the Russian era had served as one of three Russian American Company settlements. When Schieffelin visited in the 1882, Unalaska was an important regional trading center. Both the Western Fur Trading Company and its rival, the Alaska Commercial Company, operated stations there. Because of its location in the Aleutian chain and its access to open sea, it also served as an important staging point for nearly all ships bound for the Yukon or the Bering Sea. See Jacobsen, *Alaskan Voyage*, 99; Hulley, *Alaska*, 218.

Schieffelin spent a fortune on his exploration of the Alaskan wilderness. The little light-draft steam-wheel river steamer, which he commissioned to be built and named the *New Racket*, facilitated the exploration of the Yukon River and its tributaries. *Courtesy Huntington Library Photo Archives.*

brought face to face with them. Even then, sometimes with a young woman it is a question with you until speaking to her whether she is not a relation of some of the Company's employees or not.

As soon as a favorable breeze came up (which was the second day) we cleared the customs house and steered for St. Michael [at] the mouth of the Yukon River, the one we was to ascend and winter on.[15]

As I am a little ahead of my story I will go back and tell you how I got through the customs house. My steamer, being built in San Francisco and over five tons burden the *New Racket*, as I called her registering fifteen tons,

15. St. Michael was a fortified trading post on the Bering Sea about sixty miles from the Yukon estuary. The Alaska Commercial Company post at St. Michael served as the principal depot for goods destined for the Alaskan interior and was the primary jumping-off point for Yukon explorers. See Mercier, *Recollections of the Yukon*, 28.

I had to have a licensed captain, pilot and engineer.¹⁶ My engineer was all right for Old Charlie was one and had had his papers for years, not only for stationary engines on land but he had served in the Navy as a chief. All he had to do was to have them renewed. But the captain and pilot worried me. Notwithstanding my argument that the boat was altogether a private enterprise and was intended to take the place of a mule and I would just as soon think of taking out a license for a mule as to pack my grub through countries where there was no navigable streams. They said it made no difference. It had to go through the same form as if she carried thousands of tons and was made expressly for public use. So the only way out of the difficulty was for me to take out the necessary papers if I could get them. The boys all said let us get in the Yukon River and we will manage here some way between us. And I was the only one that stood a ghost of a show so it fell to me. But how was I to pass the examination? I had never sailed a vessel, didn't hardly know which end the rudder was on, and as for my steamboat experience, it was limited to being a passenger three or four times on one. I knew nothing about red lights, blue lights or any other kind of lights. But it had to be done and I got my papers, but of all the examinations that was ever passed for a steamboat captain and pilot, I guess mine beat them all. For the only question that I could answer in the affirmative was for color blindness. That I passed as well as the best. But I had my papers for Captain, Pilot and owner of the *New Racket* freight and tug steamer built for the Yukon River. It didn't make a particle of difference. We was all right then. If we met half a dozen Revenue Cutters and Customs House officers desiring to see our papers, we had them.¹⁷

---

16. In 1869 the first steamer on the Yukon was the fifty-foot *Yukon*. Schieffelin's *New Racket*, however, was the first *independent* stern-wheel steamboat to ply the waters of the river. Constructed on a rush order by San Francisco's Main-Street Ironworks, it was "a very flat little steamer" that had a single paddle wheel at the rear. See Allan, "'On the Edge of Buried Millions,'" 25; Jacobsen, *Alaskan Voyage*, 80; and Mercier, *Recollections of the Yukon*, 30.

When Schieffelin abandoned Alaska, he sold the *New Racket* to trader Al Mayo, who kept it in service at least through 1896. It was also used by other traders, including Leroy "Jack" McQuesten and Arthur Harper. See Herbert Heller, *Sourdough Sagas*, 58, 103, 120; Mercier, *Recollections of the Yukon*, 30, 35n11; and Brooks, *Blazing Alaska's Trails*, 327. See also Knutson, "Steamer New Racket."

17. In 1877 the Treasury Department began policing Alaska's waters. Until 1884, when Congress granted Alaska a civil government, the U.S. Navy and Revenue officials were the only governmental presence in Alaska. See Hunt *Alaska*, 37–45.

A few days sail from Unalaska brought us to St. Michael. About an hour after arriving, the Schooner *Leo* which had left San Francisco several days after we had and bound up to the Arctic to relieve the signal station at Point Barrows arrived.[18] On board was a party of miners bound for Golovin Bay, sixty miles from St. Michael on the north side of Norton Sound and not far from the mouth of the Bering Straits. We all went ashore—officers and miners from both vessels—and found at the station Charlie Peterson[19] whose station was one hundred and fifty miles up the river.

Henry Neumann[20] the agent there, opened some wine and beer which we all seemed to relish. Then we all went on board the *Leo* where Lieutenant Powell set out whisky and more wine which added to the already merry condition we were in.[21] It didn't take long to make us extremely jolly and set us to drinking toasts to everybody we could think of. But it lasted only a few hours [until] the *Leo* was off and we was all at work unloading and getting ready for our trip up the river. Having too much stuff for the little boat to carry, I got a bidarka from an Indian that would carry about four tons. It was made of poles about three inches thick for timbers and small sticks either bent or natural crook for ribs, and all tied with thongs cut from and covered with Walrus hides.

In a couple of days we were all ready when a storm arose which detained us for another day. We were sixty miles from the mouth of the river and our boat wouldn't stand much of a sea, nor would she steer in high wind. We had to run down the coast exposed to the wind and waves so we had to watch our chance and improve the first one, for when it commences storming in that country, it never knows when to stop.

Before the storm was fairly over, with Charlie Peterson for pilot and the cannons booming (that they always fire on the incoming or outgoing of any vessel) we steamed into the canal that separates St. Michael Island from the mainland, passing through the canal into the Bering Sea about forty miles from the mouth of the river.

18. The *Leo* was a Alaskan Trading Company schooner. See Schwatka, *Along Alaska's Great River*, 88.
19. Charlie Peterson was the manager of the upriver trading post. Eventually, he became the master of the boat *Yukon*. See Mercier, *Recollections of the Yukon*, 12, 14, 17, 68.
20. Henry Neumann was an Alaska Company agent. See Mercier, *Recollections of the Yukon*, 68.
21. Jacobsen also identifies a "Lieutenant Paul" rather than "Powell." See Jacobsen, *Alaskan Voyage*, 85.

We was going along nicely with Charlie at the wheel not withstanding there was a stiff breeze blowing. But Charlie, for years being a seafaring man, knew how to handle the boat. Jack Young, thinking it was about time some of us began to take lessons steering, went to the wheel and relieved Charlie after watching him for a few minutes. She hadn't went far before he gave the wheel a turn in the wrong way when she swung around into the wind and trough of the sea. Nor would she answer the helm although Charlie ran to and took the wheel. Still she headed and seemed determined on going out to sea, rolling and tumbling, waves splashing in and over her, looking as if we were destined to be lost. For a while it did look scaly, for she couldn't hold out long the way she was going. Charlie told us to get a tent and tie it to the steering pole and bring one end back toward the pilot house, making a kind of a jib, which just as soon as was done, she came around all right, and we ran into the mouth of a small river that put in there, where we laid for a few hours then tried it again.

We got to within four miles of the mouth of the Uphoon branch, or slough, which was to lead us into the river, the mouth of which was forty or fifty miles still farther to the southwest when we ran aground, it being near low tide.

We remained for about ten hours. At one time it didn't look much like a sea for as far as the eye could see the water only stood in pools. And Indians walked out to us from their village on the land fully five miles away.

The Yukon, being a very muddy stream has formed a bar that they say extends out into the sea sixty miles, making it as far as known impossible for sea going vessels to get any closer than St. Michael. Even our little boat, only drawing 20 inches of water, was laying on the mud high and dry, all of four miles from the nearest point of land. Made it look as if it was a question whether it was going to stay there and wait for another storm to raise the water before we went any farther.

As the sun was shining brightly, making it a very pleasant morning barring mosquitoes when the time came the tide came in and we run into the Uphoon and was safe.[22]

---

22. Both Schieffelin and Jacobsen complained bitterly of the mosquitoes in Alaska. According to Jacobsen, "The first evening and night on the Yukon River brought us a foretaste of a plague that would follow us the whole summer—the plague of mosquitoes . . . there is no defense." See Jacobsen, *Alaskan Voyage*, 91. In describing the Arctic mosquito, one prospector wrote, "He is not large enough to fry and therefore utterly useless." See Heller, *Sourdough Sagas*, 13.

The dreaded part of our journey over for we was practically in the river, the place we all always spoke of. If we can only once get in the river without any mishaps we can manage the rest. We had some little idea of what we had to contend with. But as long as we was at sea we didn't know what might turn up or where we would land. There we was sure of a landing, and if it was too rough we could camp or if the boat sunk it wasn't far to swim.

As the banks was on both sides and no chance of heading to sea, I took my first turn at the wheel of a steamboat. After sending her from one side to the other a few times (nearly running her nose into the banks) I brought her up in the center of the stream, the sweat running down both sides of my backbone. Trembling like a leaf, I held her, I thought, in good shape, but the boys all said that my course was crookeder than a lambs' tail in fly time. At all events, I managed to keep her in the water until it commenced storming and the wind raised again. When she got too much for me Charlie took her. He didn't run long until he tied up and waited for the wind to lull which took nearly twenty-four hours.

Then we run up without stopping to Andreafski Charlie's station[23] where we stopped, staying a couple of days to dry our skin boats.[24]

I forgot to mention that we had two other fellows along, one an Ethnologist, Jacobsen who came from San Francisco with us, ours being the only vessel coming up at that time, and Woolfe who claimed to be a correspondent for the *New York Herald*.[25] If he was, I never saw any of his articles. They had a small bidarka carrying themselves and what few things they had. When we was in the storm going out to sea, you could have heard Woolfe holler a mile.

---

23. Charlie Peterson operated this Alaska Commercial Company trading station when not in charge of the *Yukon*. See Schwatka, *Along Alaska's Great River*, 91.

24. According to Jacobsen, "At eleven o'clock at night we reached the trading post of the Alaska Commercial Company at Andreafski. Like most of the posts of this company it is situated where a post had been erected by the Russians before 1868 [1867] when they had possession of Alaska ... there is a well-arranged store, carrying the traditional manufactured wares used in trade with the natives for their furs and all the other goods obtained from them. There are living quarters for the trader who manages the station and also for the workmen.... Around this entire settlement there used to be a high wooden stockade as protection.... Finally there is a high cache for supplies such as dogfood, sleds, and dog harness.... Outside this settlement a number of native families usually live to enjoy the scraps from the table of the white people." See Jacobsen, *Alaskan Voyage*, 93.

25. A Norwegian by birth, Johan Adrian Jacobsen came to Alaska in 1881 to collect native artifacts for the Royal Berlin Ethnological Museum. During his two-year expedition, he sought

Afterwards, when asked what he was making so much noise for, he said that he wanted to know where we was going, something we would have liked to have known ourselves about that time.

After leaving Andreafski things went along all right, excepting one or two little mishaps. The first one the second day. After wooding up where there was a strong current—and not understanding the art of steering very well—when starting between the wind and current, the boat took a start on me and run down stream. In making the turn back again, there was a bar close to the shore and covered with water so as not to be seen. I jumped her with a full head of steam on, but fortunately it was mud and sand so that there was no damage done except to detain us for a few hours for we had to partly unload to get her off.

The next, day or two afterwards, was one of the rudders got loose and the crank pin came out, but we landed without any trouble. Keyed it up and made the mission the next day, but only stopped to wood up. The Indians, having some wood there for the *Yukon*, the Alaska Commercial Co.'s steamer, when she should come down.[26] We bought it with the understanding that they would get more for the *Yukon*.

We took on another Indian from there to show me the river—the one I had got from Andreafski claiming not to know the river any farther and it being as full of Islands and sloughs. I was afraid to undertake to hunt my way alone, so being anxious and in a hurry to get up the river, to save time I got one. Although I had to do all the steering and did all the way up (which made it pretty hard on me for we used to run about eighteen hours a day) the other boys while running could change off, but there was nobody to change with me.

---

out Indian tribes from Vancouver Island to southeast Alaska and studied the Inu and the peoples of the Yukon Valley. See Webb, *Yukon*, 103–4.

Jacobsen met H. D. Woolfe at Unalaska. A specimen collector for the Smithsonian Institution, Woolfe was indeed, from time to time, a correspondent for the *New York Herald*; later he was employed by the Alaska Commercial Company as a trader. Woolfe is credited with naming Lake Bennett for James Gordon Bennett, publisher of the *New York Herald*. According to Jacobsen, Woolfe was "a well-educated and traveled man who had been in China and has seen Alaska.... He proposed that we make the expedition together. I agreed and had no cause to regret it, for Mr. Woolfe was a pleasant, tireless helper on our journey of several thousand miles." See Jacobsen, *Alaskan Voyage*, 83 (quote), 118, 159.

26. On July 4, 1869, the *Yukon* became the first steamboat to enter the Yukon delta. It was a stern-wheeler fifty feet long and twelve feet wide, and at the time commanded by Leroy McQuesten. See Matthews, *The Yukon*, 80, 95. See also Brooks, *Blazing Alaska's Trails*, 269, 317, 417; and Webb, *Yukon*, 48–50.

At Anvik we stopped again a couple of days to dry our boats and get a new pilot. He went with us until we met the *Yukon*, where him and Woolfe turned back. We all stopped for a few hours for a visit and we got a new pilot who belonged at Nuklukayet where we was going to winter.²⁷

The next day [we] got to Nulato about noon where we dried our boats again.²⁸

Mad Dogs, Spring 1883

EDITOR'S NOTE: In addition to the overview narrative account of his trip to Alaska, Schieffelin wrote several short vignettes descriptive of life in Alaska. The first details an incident at Nulato.

It was the spring of '83 that I saw what I am telling you, at a place called Nulato, about six hundred miles up the Yukon in Alaska.²⁹

In Alaska, the dogs go mad, or have fits, which in that country they call madness.³⁰ At all events they have all the symptoms of hydrophobia, except that they don't seem to attempt to bite a person, but snap and bite at anything else that comes in their way. Everything that they bite has the same disease whether it is hydrophobia or what not.

It was while traveling with dogs, on my way home after being gone all winter that I stopped there a couple of days to rest. I had something like three hundred miles farther to go. The night before I started one of the Indian's dogs went mad and as usual, the Indians undertook to kill it with clubs. After

---

27. Nuklukayet, an outpost of the Alaska Commercial Company, was located at the confluence of the Yukon and Tanana Rivers, seventeen miles below the present town of Tanana. The area had long been used by indigenous peoples for annual gatherings. See Heller, *Sourdough Sagas*, 119, and Mercier, *Recollections of the Yukon*, 30.

28. Located about six hundred miles upstream from St. Michael, Nulato was established as a post in 1842 by the Russian-American Company. Though destroyed several times by Indians, by the time Schieffelin arrived, the fear of Indian attack was virtually non-existent. Both the Alaska Commercial Company and the rival Western Fur Company maintained outposts there. See Hunt, *Alaska*, 52.

29. Schieffelin, episode fourteen (untitled), *Memoirs*.

30. Three principal breeds of dog were used for sleds: the husky, a descendant of the red wolf of the McKenzie River valley; the malamute, the largest, a descendent of the grey wolf of the Arctic; and the Siberian, whose ancestor is the Arctic wolf of the Siberian Coast. See Young, *Adventures in Alaska*, 80.

For posterity's sake, Schieffelin's companion, Charles O. Farciot, photographically documented the party's gold-prospecting trip. This is a picture of Schieffelin with his trusted dog team. *Courtesy Huntington Library Photo Archives.*

about ten minutes of quite an exciting time, and to them apparently having a good deal of sport, they succeeded or supposed they had. To all appearances the dog was dead. He lay stretched out just under the bank of the river and close to the road that I was to take in the morning, the blood running out of his nose, stiff as a post.

In the morning when I started on passing where he laid, I noticed that he was curled up and breathing and looked like a dog asleep, but I went on not stopping to examine him or give him any further thought than that he must be tough to stand such a beating and live for it was the intention of the Indians to kill him. They seemed to try hard enough for when they succeeded in knocking him down after several hard licks, mostly in the neighborhood of his head, they then struck him three or four times in his forehead hard enough I thought to kill most anything, causing him to stiffen and tremble like any animal that is killed by having their skull crushed.

The following fall, when I left the country for San Francisco, [I met] the trader Captain Waldren[31] who was at Nulato at the time I was then at St. Michael, the port for that country where I was a few days before the revenue cutter arrived that brought me to San Francisco. We immediately got to talking about the circumstances when he told me that that dog got well, and was alive and hearty as any dog in the country. He said that that was not the only case where they had got well under similar treatment, that he himself knew of quite a number of cases, and he said that it was usually conceded by the Indians that if the dogs lived (although they tried to kill them) they got well. The white men usually shoot them, not only to relieve them of their sufferings which seems to be intense, but to protect them from the cruelty of the Indians.

I had two that went mad. One I didn't see but the other was where I was at a station during the winter. It was a dog that I had but a day or two previous bought from an Indian and being a stranger to him had him chained to keep him from running away. I never saw anything like him—his eyes was green and the slime running from his mouth, not froth or anything white and of a frothy substance, but the nearest comparison I can make is the white of an egg. It resembled that more than anything else and a profusion of it. Although it was middle winter and very cold, he seemed to be in a very high fever and constantly struggled with all his might to break loose. It paid but little attention to anybody around him. He acted as though he was in the most intense agony, standing on his hind feet, howling, yelping, and snapping continually, but no barking. As soon as I saw him I had him shot.

What effect the beating given by the Indians has on them I don't know. Consequently offer no suggestions, but I would like some scientist to try and see if he could offer any. For instance what would Pasteur think of such a treatment for hydrophobia?

After leaving Nulato, about forty miles above, we ran over something in the river that knocked one of the rudders off, losing the cap to the lower box holding it.[32] But we made a new one out of wood and went on losing

---

31. For Arthur Treadwell Walden's adventures in Alaska, see his book *A Dog-Puncher on the Yukon* (New York: Houghton Mifflin, 1928).
32. Schieffelin's *Memoirs* continues.

While Schieffelin and his brother Eff explored up the Yukon, several other members of his party built a comfortable cabin at Nuklukayet Station, a trading post of the Alaska Commercial Company. Trading stations served as important social centers for both traders and the indigenous peoples throughout the long winter months. *Courtesy Huntington Library Photo Archives.*

about an hour's time. Finally we rounded a point and there was our station in sight.[33] The Indians commenced firing their guns, rather a different welcome from what we had been used to from Indians. It might have frightened us had we not known that it was a friendly greeting. There at Nuklukayet we unloaded our stuff putting it in one end of the trader's store, it being built with three rooms to it. Al Mayo, the trader kindly gave us one end of it.[34]

While the boys was building ourselves a good, comfortable cabin for the winter at Nuklukayet a station of the Alaska Commercial Company (Al Mayo being treasurer there), my brother and I took a canoe—a three-hole bidarka

---

33. According to Jacobsen, the party arrived at Nuklukayet on Sunday, August 27, "happy that the towing job was over." Nuklukayet, an Alaska Commercial Company station, is where Schieffelin decided to set up a winter camp. See Jacobsen, *Alaskan Voyage*, 103.

34. Schieffelin first met Al Mayo, a Kentucky-born prospector, aboard the *Yukon*. Mayo was one of the first prospectors in the Yukon Territory. Like other early whites

such as the Indians living along the coast use—and started out prospecting to see what we could find before housing up for the winter.³⁵

The Indians around where we were then, are the Ingliks and all use birch bark canoes, while the Indians on the coast all use the skin boats. We had one of these. The three-hole bidarka was a very good boat for prospecting in that country for two men can sit (one in the bow, and the other in the stern and in the middle hole carry provisions) putting a piece of canvas over to keep them perfectly dry. In a country like that, where there is so much rain, this arrangement is very convenient.³⁶

The weather then was pleasant, only an occasional rain and no mosquitoes but millions of gnats which bothered some of the boys as bad as the mosquitoes because they could devise no way of protecting themselves. We found nothing to attract our notice until the eighth of October when about 60 miles above our winter quarters, in a gulch in the lower ramparts of the Yukon that emptied into a creek about ten miles from the river, we found quite a prospect.³⁷

---

in the area, he first tried his hand at prospecting before becoming an agent for the Alaska Commercial Company in 1875. He remained in Alaska for decades and in 1897 became mayor of Rampart. See Jacobsen, *Alaska Voyage*, 99; Mercier, *Recollections of the Yukon*, 3–68; Wharton, *The Alaska Gold Rush*, 17–22, and Webb, *Yukon*, 60.

35. In another account Schieffelin noted, "[We] arrived at our destination in the latter part of August. Here, while some of our party were building winter quarters, two of us were prospecting. As soon as we got there with our little boat we towed the logs for the purpose of building a cabin and also towed up a lot of wood for the winter. We built ourselves a good, comfortable log cabin, and there spent the winter. While the boys were building this cabin, my brother and I took a canoe and started out prospecting." See Bancroft, "Alaska," 2–3.

36. For the above two paragraphs (beginning "While the boys" . . . and ending with "this arrangement is very convenient."), see Bancroft, "Alaska," 3. The paragraphs hereafter are from Schieffelin, episode fourteen (untitled), *Memoirs*.

37. The Bancroft manuscript version describes the discovery of the small waterway that came to be named "Schieffelin Creek" slightly differently: "We prospected without finding anything until about the first of October, when about 60 miles above our winter quarters, in what is known as the Lower Ramparts of the Yukon, on a stream emptying into the main river about ten miles from its mouth. On the 7th of October we got a fair prospect, of the coarse kind of gold, but it was too late to do anything, although the next day, the 8th, we took a pair of blankets and a little food and went up there and stayed there that night, after prospecting as much as we could that afternoon, that night and the next morning." See Bancroft, "Alaska," 3–4. McQuesten recalls, "there was big excitement over prospects that Sheffelling [sic] had found the fall before [on the creek that] he had named 'Maybeso Gulch.' We were showed some of the gold; it was very coarse pieces and some of it was very black. . . . They were greatly disappointed in the gulch—they could not find anything that would pay wages." See McQuesten, *Recollections of Leroy N. McQuesten*, 12.

Taking a pair of blankets on our backs and killing some pheasants that were very numerous there for something to eat, we camped one night at the gulch. Eff, not being used to such ways, fared pretty hard. He had to set up nearly all night and keep fire to keep from freezing for try as he would, he could not roll up and stay so in his blankets. We didn't have but a pair apiece and consequently we could not improvise a bed of any kind. It was very cold which caused him a great deal of suffering.

The next day we went to winter quarters, drifting down the river with the anchor-ice.[38] We stopped about forty miles down at Walker's who went down with us the next morning to our winter quarters but we had to leave our boat on an Island, the slough between it and where the station being frozen.[39] The next day, the eleventh, we went skating. That night we tapped the keg of whisky and got a little merry, but not drunk because we didn't have it there for that purpose.

When we came to open our supplies for winter use we found fully half of them damaged if not more. Some of them so bad that we had to throw them away—utterly unfit for use. When I gave the order at Albert Man's and Company in San Francisco, I told them particular that I want nothing but that was good and first class articles, for we was going where it was impossible to getting anything. What we didn't have we would have to do without, and that was the way they served us. Fortunately, we had a large assortment or we would have fared poorly.[40]

I think it was about the fifth of November that Mayo and myself, with dogs and sleds and an Indian apiece, started for St. Michael [about a four-

---

38. "The ice was making in the river then; the banks along the sides of the river for some little distance out into the river were frozen, but the current was still open, but filled with what is called 'anchor ice,' that is, floating ice." See Bancroft, "Alaska," 4.

39. "When we got to our winter quarters, the ice was frozen so we could not get our boat within a mile of the house; we left it on the island, and Mayo and Jack Young got into a canoe with a couple of Indians, and breaking a way across the slough which separated this island from the main land where the cabin was built, with a club, they brought us over; the day after we hitched up the dogs and went over across this slough with dogs and sleds, and loaded our bidarka upon the sled and brought it over." See Bancroft, "Alaska," 4.

40. With part of the twenty thousand dollars that Schieffelin spent outfitting his party, he had purchased picks and shovels, axes and other mining tools, heavy woolen clothing, camp equipage, rifles, and revolvers, as well as seven thousand pounds of flour, five hundred pounds of coffee, and, to help keep their innards warm, ten gallons of whiskey and five of brandy. See Allan, "'On the Edge of Buried Millions,'" 25–26.

hundred-mile trip]. He was going for some supplies needed in his trade and I for dogs. Calculating to be back before Christmas but didn't get back until the 14th of April.

When we left the station it was snowing and rather warm. The snow was soft so that it was very slow getting along, having to continually wear snow shoes, instead of riding the sled as I had always supposed. (I had so often seen traveling in the arctic regions illustrated as a man setting in a sled wrapped up with furs and dogs going like the wind—but my experience during that winter was far different). I found out that to ride with dogs and sleds it was as necessary to have a good road as anything else. Without it, instead of riding the sled it would be the snow shoes and walk behind like plowing and steer the sled, there being a pair of handles behind for that purpose.

The first day we only made about twelve miles and we had three Indians ahead to break the road (the third one going down the river with us to Novikakit about seventy miles. That night it stormed all night and the next day also [so] that we laid there. The next day, having no more feed for our dogs and still storming, and fearing a rain which would make traveling impossible, we returned to Novikakit to wait until the weather should settle and get more feed for the dogs. There we stayed until about the 20th of December, then started again. All the way to Nulato—which took us about two weeks—the snow was very soft. So much so that it was necessary to break the road to get along at all. A good deal of the way, on the ice under the snow was water, which [we] would often sink into, then we would have to stop and clean the ice from the runners as well as our snow shoes. This would often occur every few minutes and some of it appeared as if we was doing nothing else all day but cleaning our sleds and snow shoes.

But after leaving Nulato (where we stopped to rest a few days), we had a good road for about forty miles down the river to Kaltag where we left the river. And for the portage across to Norton Sound—about sixty miles where the same slow traveling as previously experienced—was gone over again only there was no water. There was several sleds and quite a number of Indians, all on snow shoes and those that wasn't steering a sled was ahead breaking a road so that those behind had a fair road.

We got to the Utukok River, twenty miles from the Sound. At Ivans I met a Malamute Indian. His wife was the prettiest woman I think I ever saw, only she lacked that intelligent look that cultivation gives. But aside from that, she was beautiful. A half-breed, her mother being a Malamute woman and her father a Russian (one of the ignorant ones, I suppose a serf—at least he was about the same as an Indian). I noticed wherever I went that the half-breeds were handsome and some of the children of the traders were pictures.

During our trip across the portage it was the coldest weather we had during the winter. The morning we left Nulato the thermometer that Captain Waldron, the trader, had belonging to the Signal Service registered 55 degrees below zero. It seemed to me to be much colder on the portage a day or two after than it was that morning. It would seem almost impossible to camp out and live in such cold weather, but by preparing camp every night a person can be very comfortable. The mode of preparation is to clear the snow away making a pit as it were. Then spread some spruce boughs on the ground and stick some poles in the banks of snow made by clearing away, always taking care to have the wind right so that the smoke and sparks will pass away. Then, around poles which should stand at about an angle of forty-five degrees, stretch a tarpaulin or sheet of any kind. And put some more boughs around the sides so as to break the wind off, should there be any. Then building the fire in front throws its heat up against the sheet and is reflected back, keeping you warm on both sides. With a few furs to lay on it is a very comfortable camp.

Finally we arrived at St. Michael a little before Christmas where I stayed something like a month, having a pleasant time as far as possible under the circumstances. With the two traders Lorenz and Neumann, and the signal officer Leavitt as my companions, as well as Mrs. Lorenz who done all she could to make my stay as pleasant as possible.[41]

We remained there at the station until the forepart of December, when we started out, and with a good deal of hard work got down to St. Michael.[42] We ate our Christmas dinner there. There I stopped for a short time, and procured a good team as I had no dogs until this time. I got a good

---

41. Moise Lorenz served as the commissary agent for the Alaska Commercial Company. His wife, "a beautiful and refined American woman from Maine," brought with her to Alaska all the trappings of culture—wallpaper, carpets, and even canaries in gilded cages could be found at their home. See Webb, *Yukon*, 68.

42. Bancroft, "Alaska," 4–5.

team of dogs and a sled there, and then started for the Kuskokwim. I had an Ingalik Indian with me who could understand some English, but he could not speak it beyond a few words. We could hold no conversation, but still I could make him understand everything that was necessary.

We started out to go to the Kuskokwim in company with a trader [Ned Lind] in the employ of the opposition trading company (the Western Fur and Trading Company). We went across the country to Anvik Station and from there down the Yukon River 50 miles to the Indian village called Paimiut.

Native Ways, 1882–83[43]

Just about as thin a dress as I ever saw, was on some Indians at a village in Alaska called Paimiut.

In the winter of '82 or '83 I was in that country and traveled a large part of it with dogs and a sled, and sometimes was in company of some of the traders that was living at different stations trading in furs among the Indians.

At this time of which I am speaking I and Ned Lind left St. Michael on Norton Sound for Kolmankofsky, a station on the Kuskokwim River about three hundred mile in an easterly direction. We went across the country from St. Michael to and down the Anvik River, to Anvik station at the mouth and where the Anvik empties into the Yukon, about one hundred and fifty miles, then down the Yukon about fifty miles to Paimiut and across the country to about one hundred miles to Kolmankofsky.

The night we stopped at Paimiute the Indians were having an Egruska, or dance, and there was quite a number from other villages there together with those belonging there. It being a village of probably fifteen hundred, made quite a crowd so that their kashim was as full as it would hold conveniently. Still they made room for us in one corner, it being the place where they always take strangers.

The kashim is built partially underground, like all their winter habitations anywhere near the coast. It is entered by a tunnel, and a hole cut in the center of the floor, which is composed of spruce boards split and hewed out. A portion in the center (around the hole) is loose so as to be removed when necessary

---

43. Schieffelin, episode fifteen (untitled), *Memoirs*.

to build a fire to heat it up which is about once a week. The smoke escapes through a hole in the center of the roof which is afterwards covered with a transparent covering made of seal gut, the only means of light during the day. At night using seal oil which is held in a stove basin nearly as large as an ordinary wash bowl using for a wick, moss. It stands on a post about three feet from the floor. It can be imagined what it must be when lit, that rancid oil and green moss burning together sending up a stream of black smoke. Where there is two or three of them burning at once, and the room is full of Indians of all sexes and ages without the least ventilation and that for hours at a time [you can imagine] what it must be like.

The inside on average is thirty or forty feet square. The walls built out of poles about ten feet long, split in half and set end on end with the cracks checked with moss. The roof has four sides to it and coming to a center where the hole is for the smoke to go out. The whole thing is underground. The dirt used in excavating for to put the house is put on the roof. A little house over the door at the entrance of the tunnel keeps the snow from piling upon it. There is a plank about two feet wide around the inside of the house about three feet from the floor. I believe completes the description of what a kashim or barabara is, for they are both built alike. One is for a hall or whatever you are a mind to call it—it is where all councils and dances or anything of that description is held; the other is for habitation or a house where different families live.

When we arrived they were not dancing but that evening they asked Ned if we had any objection to their continuing their dance, and he asked me. I told Ned to let them go ahead. So at it they went and after three or four hours they concluded to have supper, or at least I supposed it was supper.

At all events the women brought in Kantags (a dish of all sizes made out of birch, a hoop something like a cheese hook with a bottom to it) full of all kinds of stinking stuff and placed them on the floor in front of some old men that I had noticed lounging on the plank. I thought they were naked, and they were as my judgement on such things goes. After the women had brought in all they intended to and all was seated, girls and boys men and women, old and young, three of these old fellows that was naked got up and commenced handing the eatables around doing the honors as it were. Coming to us they offered or asked if we would have anything. I declined; Ned took something, I have forgotten what.

Then I saw what I thought my eyes deceived me in. The old man had drawn the foreskin over the head of his penis and tied a string around it. It was the only semblance of clothing, if that might be called, such that he had on. I asked Ned what the devil is that string there for.

"Why," he says, "they don't consider they are naked when that is on. Don't you see it hides the head of it? And that, according to their custom, is what they call naked. Once one of those others over there that is naked, gets up, if he hasn't got the string on, he will cover it with his hand or something."

That was true, for a few minutes before I noticed one of them walking across the floor holding his hand over it. And that with the indifference in which those other three was walking around, together with Ned's explanation, him being well acquainted with their habits having, until a year or two previously, lived within a few miles of that village for a few years and was conversant with their habits. At least that was his explanation and from what I saw I believe it.

It was the littlest and lightest evening dress that I ever saw—a piece of string about the size of an ordinary shoe string and about a foot long. But it shows what custom does. There wasn't a soul paying any attention to them but me, anymore than if they had all the clothes they could carry on them.

Ned and I left Paimiut and thence went across the country to the Kuskokwim, to Sipray station, intending to go up the Kuskokwim River and over upon the Tanana [River] and down the Tanana home.[44] But it was such stormy weather that winter, and there was no feed for the dogs up the river (it was said that the Indians had no fish). I therefore did not make the trip, for fear of losing my dogs. I went across to the Yukon River, by Nulato, home, getting back about the 14th of April.

While I was gone and after Mayo had returned in the winter, the boys got his dogs and Eff, Sauerbrey and Jack took Mayo's dogs and some provisions and went up there to this [Maybeso] Gulch, and built a little sort of tent out of spruce poles, covered with brush and snow, and tried to do some prospecting. They had gotten tired of lying around the cabin all winter. They said that this gulch where we found this prospect, though full of springs of some kind of

---

44. Bancroft, "Alaska," 5–10.

mineral water, from some cause did not freeze during the winter. At all events, when I got back in April (it on the 19th) we took Mayo's team and the one that I brought, and loaded them with provisions and went up there, taking a couple of Indians with us whom we sent back with the teams after we got up there. We left Charlie O. Farciot to look after our provisions at the winter quarters and also to guard the little steamer.

After getting up to this gulch, although it was April and the country was covered with snow and still frozen up, we went to prospecting right along in this gulch, where it was still open, and quite a long cut two or three feet wide where it was frozen. It must have been 150 or 200 feet long, down into the bed-rock, and we got some very good prospects of coarse gold. The gold was black, due I suppose to iron rust. It was very smooth and quite heavy gold when this black was rubbed off of it. We stayed there at that camp until June, after the ice went out of the river. It went out that year on the 29th of May.

Then, with our three-hole bidarka and a canvas boat which Jack Young had made during the winter (something similar to one of the [Athabascan] birch bark canoes, which he had taken as a pattern) he made a frame and covered it with a portion of canvas we had brought with us, which we found we would have no other use for. He painted this boat and made a very handy craft of it. It was light so that a man could carry it anywhere and it was still strong enough for service. With that and the bidarka we started out prospecting on the small streams which run into the river there, not using our little steamer.

We remained in that country until August, when we came to the conclusion that the most of the country was frozen up, wherever we went, so much so that it would be almost impossible to do much work. We did not see anything very encouraging. If we found anything it would have to be immensely rich in order to pay. The boys that were with me had nothing to depend upon excepting their labor; they had no homes, and time was flying and they had to make homes for themselves. There were other countries in which there were mines to be discovered and where the climate was not so terribly rigorous as it was up there. And so we thought, taking everything into consideration, that we had better come back; that the chances of our discovering anything in that country, with the limited time that we would have to work and the inconveniences that would have to be overcome, that it would have to be so immensely rich in order to make it worth while and that we might stay there forever and not accomplish anything. We therefore came back.

If I had been alone, I should have remained another winter at least, and probably two winters. Although I had borne all the expenses, furnishing provisions, clothes, and everything, the men who were with me were losing their time, and I did not wish to hold them, although they had agreed to stay with me as long as I wished them to. They were willing to remain, but as we did not find anything, and they were not deriving any profit for the venture, we disbanded the company.[45]

We had drawn up an article, when we started, a kind of agreement, under which I was to bear all expenses, to take the other boys up there, furnish them with provisions, tools, clothes and everything else; in fact to bear all expenses of the trip, bring them back to San Francisco and give them $200 or $300 apiece, so that they could go anywhere, to any part of the country they wanted to in case they did not find anything on this trip. We were to be gone for three years, unless by mutual agreement we disbanded. This we did, and disbanded the company there so that in case any one of us wanted to stay, he could do so.

We all signed an agreement to disband at St. Michael, and Charley Farciot remained two or three years afterwards.[46] The little steamer that we brought with us, we sold to traders there, Al Mayo, Jack McQuesten[47] and Arthur Harper.[48] In making the sale we reserved the right to go down the river with her.[49]

---

45. This paragraph is from Bancroft, "Edward Schieffelin: The Discoverer of Tombstone," 8–9.

46. Farciot remained in the employ of the firm of McQuesten, Harper and Mayo as engineer and stayed in the region until 1885. Eventually he moved to California, where he was employed as an engineer in Richard Gird and H. T. Oxnard's Chino, California, sugar beet factory. He worked there until his death on October 27, 1891. See Webb, *Yukon*, 71, and Rowe, "Arizona Views of Charles O. Farciot," 377.

47. Leroy N. "Jack" McQuesten (1836–1909) was born in Litchfield, New Hampshire, and came to British Columbia in 1836. Considered the "Father of the Yukon" by some, he was reportedly one of the most trusted and popular men on the river. He arrived in Alaska in 1873 and, like Harper, prospected unsuccessfully for gold. See Hulley, *Alaska*, 222; David Wharton, *The Alaska Gold Rush*, 17–20.

48. Arthur Harper (1835–97) was born in County Antrim, Ireland, in 1835. A consummate prospector, he arrived in California and gradually worked his way northward. In 1872 he was one of the first men to prospect for gold on the Yukon River and is credited as the first to actually find gold on the lower reaches of the Forty Mile, Stewart, and Tanana Rivers. Though he found gold in many streams, he never found a major strike. See Sherwood, *Alaska and Its History*, 354; Hulley, *Alaska*, 222; Mathews, *The Yukon*, 86; Brooks, *Blazing Alaska's Trails*, 312–14; and Webb, *Yukon*, 59–60.

49. When Schieffelin decided to sell, he could not have found more eager prospects for the *New Racket* than Harper, McQuesten, and Mayo, all of whom wanted to quit the Alaska

When the Alaska Commercial Company's steamer came up the river with goods for different trading posts, we learned that the revenue cutter *Corwin* was up in the Arctic, and would have to put in at St. Michael for coal on her return to San Francisco. Therefore, if we could get down to St. Michael in time, the cutter would bring us down to San Francisco. This was the reason we reserved the right when selling the steamer to come down to St. Michael, and also to go back in case we missed the revenue cutter. Al Mayo came down with us, so as to take the little steamer *New Racket* back. We reached St. Michael just a few days before the cutter, which brought us to San Francisco, leaving St. Michael the latter part of September. We left not finding anything of importance.[50]

There was one result of the trip, however, it satisfied us, as we were always anxious to know about the country, and always wanted to see it. We left here more or less impregnated with the malaria that is contracted in Arizona, in fact two of our party had the malaria pretty severely and so much so that physicians advised them not to take any such a trip as that before taking a course of medicine and preparing themselves thoroughly, as it was suicide to go off into a country as far as that away from civilization and supplies in the condition they were in. But they paid no attention to what the physicians said, and went on, abandoning all medicine. We took a little medicine chest along with us, but it was never used during the whole trip for any of our party. The only use it was put to was during the winter when a squaw who was suffering with typhoid pneumonia, so the boys called it, was thrown out to die. The boys took her into the cabin and gave her paregoric and cured her, or at least she got well in a very short time.

These boys that started up there with the malarial fever, in fact the whole party, came back as strong and healthy, stout, robust a set of men as anybody

---

Commercial Company and start their own business. After purchasing the boat, they used it as a supply boat and to mine gold on the Stewart River in 1886. See Mercier, *Recollections of the Yukon*, 35n11; Allan, "'On the Edge of Buried Millions,'" 33.

The three former prospectors spent over twenty years together, in partnership with each other in various endeavors. McQuesten operated the boat the *Yukon* and traded at Fort Yukon while Harper and Mayo traded at Fort Reliance; all three had an interest in the trading post at Nuklukayet. See Hulley, *Alaska*, 223; Heller, *Sourdough Sagas*, 123.

50. This line is from Bancroft, "Ed Schieffelin: Discoverer of Tombstone," 8–9; the materials that follow are from "Edward Schieffelin's Trip to Alaska," 10.

Schieffelin never seemed to take much to married life; starry nights in the desert called to him instead. Shortly after they wed, Mary Brown settled into an Alameda mansion while Schieffelin launched a series of prospecting trips that took him to far-flung destinations in the American West. *Author's collection.*

saw, no more like the same party that left San Francisco, at least the sick members, than if they were not the same men. Still, during the summer season while we were there, there was so much rain and the country was so covered with brush and this spruce timber, that we were constantly wet. Whenever we left our boats, unless we waded up the gulches or cracks we would have to carry a hatchet with us and one of us go ahead to cut the way through the brush, while the other carried the tools. So, after spending two summers and a winter in the country we came back satisfied that we wanted no more of Alaska.

EDITOR'S NOTE: When Schieffelin and his party arrived back in San Francisco, they were greeted by a bevy of reporters. Though he expressed pleasure with the general health of his compatriots (the frigid cold of the Alaska winter seemed to have cured several in his party of the debilitating effects of malarial fever that they had originally contracted in Arizona), he had little else positive to say about Alaska. "The country will never amount to anything" he asserted. When one reporter asked whether he would venture into the region again, the prospector answered in no uncertain terms, "No sir; nor would I advise anybody to go there—not for any purpose unless he wanted to die of starvation."[51]

On my return from Alaska, about two months afterward, I was married, in the fall of 1883 to Mrs. M. E. Brown, a native of Virginia, at that time residing in San Francisco.[52] The marriage took place at La Junta, Colorado. I had been in New Mexico, and as we neither of us desired a pretentious wedding, I wrote my fiancée to meet me at Albuquerque, which she did, and we went to La Junta, and were married there. We spent part of the winter in Salt Lake City, then came to Alameda in the spring of 1884 and I bought my present home there.

---

51. Quote, see Allan, "On the Edge of Buried Millions," 34.
52. From Bancroft, "Edward Schieffelin: The Discoverer of Tombstone," 7–9. Schieffelin's romance and life with Mary Brown is best chronicled in Schieffelin and Butler, *Destination Tombstone*, 8–11.

In the winter of '84, I removed with my wife to Los Angeles and resided upon the place I had bought (a fine residence and orchard of A. B. Chapman Esq. in East Los Angeles for $21,000),[53] building a smaller place for my mother. During the summer of 1885, my brother [Albert] who had been interested in Tombstone with me, came there, and remained until he died, in October 1885. He had contracted consumption and died of this disease.

[At present,] one of my brothers is still in Oregon, engaged in farming. My sisters reside in Los Angeles; one sister, [Elizabeth Jane,] is married to R. C. Guirado, a real estate dealer there, and the other [Charlotte] is the wife of Mr. Ed. Dunham, proprietor of the Nadeau House, Los Angeles.

Since [1885] I have made two or three prospecting trips through different parts of the country, but have made no discoveries of importance.

---

53. The parenthetical statement is from "Ed's New Home," in Green, "Scrapbook," 4.

→ 7 ←

Ever a Prospector, 1883–1897

THE STORY OF ED SCHIEFFELIN'S LIFE as told in his memoirs and oral histories ends in the mid-1880s, but his prospecting expeditions did not. Just prior to leaving on his Alaska adventure, he had told an acquaintance that if the Yukon trip failed to pan out, he would cross over to Asia and travel down the coast in search of gold.[1] Though he never visited Asia or Africa (as he told reporters he hoped to do upon his return from Alaska), or Europe for that matter, throughout much of the 1880s he did travel extensively in the United States.

Schieffelin became a national celebrity after the Tombstone discovery and made a "grand tour" of the East, visiting New York, Washington, D.C., Philadelphia, and Chicago. He stayed in the finest hotels, dined at expensive restaurants, and engaged presidents and foreign dignitaries in conversation. Additionally, because of his penchant to give money away indiscriminately to hospitals, orphanages, and charity houses, their agents followed him from city to city. But Schieffelin soon grew tired of the life of a dandy philanthropist and returned to his first and only true love—prospecting.[2]

In 1884, after being married only a few short months, Schieffelin launched a well-equipped expedition "into the Papago country" of Arizona. With a custom-built wagon that carried his prospecting tools, he also visited several mining districts in California and others in Arizona. He then returned home, his desire for adventure again satisfied.[3]

In September 1896, after a brief stay at home in San Francisco, Schieffelin felt the urge to prospect yet again. He bought another fine outfit consisting of

---

1. "Finding His Tombstone," in Neta Green, "Scrapbook," 22.
2. For Schieffelin's forays into the eastern metropolises, see Burns, *Tombstone*, 22. See also Turner, "Ed Schieffelin," 17.
3. Rice, "The Schieffelin Brothers Ed and Albert."

a wagon, several mules, tools, cooking utensils, and provisions sufficient for several months. This time, he headed north to Oregon, wandering familiar roads in the mountains of Jackson County, the country of his childhood. After arriving home in grand style, he visited friends, fished, and sat for portraits. Following a few weeks of relaxation, he prospected in remote areas of California and Nevada, but eventually returned to the vicinity of Roseburg, Oregon. His companion on these journeys was an eighteen-year-old youth, Charlie Warren, whom Schieffelin met in a blacksmith's shop in Oregon. With Warren serving as teamster and camp maker, once again Ed Schieffelin was doing what he loved best—prospecting.[4]

In September 1896 Schieffelin briefly and abruptly returned to Alameda, California, to make out his last will and testament—perhaps he felt heart palpitations, signaling a possible early death. Upon his return to Oregon in 1897, though, he bought a small ranch near his brothers Effingham and Jacob, just outside Woodville (today known as Rogue River). From here he continued prospecting for gold and silver in the Canyonville area.

Ed then decided to try his luck in Douglas County and once again recruited young Charlie Warren to accompany him. They traveled through Grants Pass and camped on Days Creek, where they took up residence in an abandoned cabin on a ridge above the creek some twenty miles from Canyonville, not too far from present day Medford, Oregon.[5] There, on May 12, 1897, Schieffelin died. He was only forty-nine years old.[6]

Though some have suggested Schieffelin's death may have resulted from foul play, it is far more plausible that he simply died prematurely of heart failure. A few days before his death, he had told an old acquaintance that his rheumatism and neuralgia had flared up again, that he was not feeling well, and he planned to stay inside until he felt a little better. When the body was

---

4. Turner, "Ed Schieffelin," 17.
5. The area associated with Schieffelin's last prospecting is in the vicinity of the canyons along the South Umpqua River, near Days Creek, seven miles east of Canyonville, Oregon, just off Highway 227 (Interstate 5, exit 98). To reach the site of Schieffelin's last homestead cabin, travel to the intersection of Schieffelin Gulch Road and Highway 99. From Interstate 5, take exit 48, cross over the Depot Street Bridge, then turn left and continue for one mile. Schieffelin Gulch Road is on the right. See "Most Miners Hope to Strike it Rich," *Medford (Ore.) Mail Tribune*, May 16, 2010.
6. Schieffelin's body was found on May 14, 1897. Because he died alone, the actual death date is conjectural. See Underhill, "The Tombstone Discovery," 40n6.

Schieffelin grave and monument, installed circa 1900. *Author's collection.*

viewed by Canyonville sheriff Alex Orme, Schieffelin lay face down, full length on the floor near the doorway of his cabin. Apparently, he had started cooking a meal and had sat down in the doorway to read a book while the food cooked. When the sheriff studied the scene, he noticed that the beans Schieffelin had been cooking were boiled down to a charred mass, his bread was burned black, and the stove was stone cold. The logical conclusion: Schieffelin had died suddenly of natural causes and without warning.[7]

7. Other accounts claim he was at a table breaking ore with a hammer. See Turner, "Ed Schieffelin," 17. For contemporary accounts of his death, see *Los Angeles Times*, May 16, 1897; *Tucson Arizona Daily Star*, May 18, 1897; and *Tucson (Ariz.) Citizen*, May 22, 1897.

Near the body was a small bag of quartz. It was assayed at over two thousand dollars a ton, making it as rich a prospect, if not richer than, the Tombstone discovery. Only a few days before his death, he had written his mother, telling her of his discovery and that he considered it another great strike. "I have found stuff here in Oregon that will make Tombstone look like salt (that is a salted and worthless mine). This is GOLD."[8]

Shortly after its discovery, Schieffelin's body was wrapped in a blue woolen blanket and temporarily buried under a tree not far from the doorway of the cabin. When Charlie Warren returned to the cabin a few days later, after an extended visit to his parents on the Rogue River, he told the sheriff that Schieffelin had two woolen blankets in the cabin and that one—the red one—was missing. Warren surmised that Schieffelin probably left the red blanket at a camp and that the new strike was probably close to it.

News of Schieffelin's death and his discovery of a rich prospect touched off a brief gold rush in that section of Oregon. Since the late 1800s many have tried to find Schieffelin's last and perhaps richest strike, but for over a century the lost "Red Blanket Mine" in the vicinity of Canyonville, Oregon, has eluded discovery.[9]

Schieffelin's body was disinterred and then shipped to Tombstone. On May 23, 1897, a bright Sunday, the bells tolled for him at high noon, when his

---

8. Ed Schieffelin's letter to his mother as quoted in "Tombstone Ed's Treasure," *True* (January 1945): 46–47, 90, 92, 95 (quote). Other accounts claim that Schieffelin's journal lay near the body and on the last page he had scrawled, "Struck it rich again, by God." See "Most Miners Hope to Strike It Rich; Ed Schieffelin Managed to Do It Twice," *Medford (Ore.) Mail Tribune*, May 16, 2010.

9. Primary-source accounts that document details relating to Schieffelin's death are sketchy and not entirely trustworthy. See Anon., "Ed Schieffelin's Lonely Death," *Ashland (Ore.) Tidings*, May 17, 1897; and Dan Green, "Comment Upon Men and Issues," *Los Angeles Independent Review*, in Green, "Scrapbook," 49.

By contrast, the literature relating to the Red Blanket Mine is plentiful, especially in newspapers and treasure-hunter-related publications. See Carl Gohas, "Schieffelin Struck It Again, but Where's His Red Blanket Mine?" in *Portland (Ore.) Reporter*, October 23, 1963, 11; "Missing Red Blanket may Be Clue to Ed Schieffelin's Lost Mine in Canyon Creek Region," *Roseburg (Ore.) News-Review*, April 15, 1948; Ruby E. Hult, "Ed Schieffelin's Lost Oregon Gold Mine," in *Treasure Hunting Northwest* (Portland, Ore.: Binfords and Mort, 1971), 155–59. For a fascinating account of one prospector's eight-year quest to find Schieffelin's lost mine see Jess Minckler, "Tombstone Ed's Treasure as Told to Stewart Holbrook," Arizona Historical Society Collection, Tucson, Arizona. For more literature on the mythic lost mine, see Probert, *Lost Mines and Buried Treasures of the West*, 391–92.

friends and family conducted a brief funeral service. It was the largest in the town's history. Stores were closed, flags suspended from the spires of public buildings flew at half mast, and Schieffelin Hall was draped in black crepe. In attendance were his wife, mother, and brother Charles, as well as friends and hundreds of passing acquaintances.

Following the service, Schieffelin's friends carried his body a short distance out of town where he was interred under big boulders (now a large rock cairn), thus fulfilling Schieffelin's final request: "It is my wish, if convenient to be buried in the dress of a prospector, my pick and canteen with me, on top of the granite hills about three miles westerly from the City of Tombstone, Arizona and that a monument such as prospectors build when locating a mining claim be built over my grave, and no other slab or monument erected. And I request that none of my friends wear crepe. Under no circumstances do I want to be buried in any graveyard or cemetery."[10]

There, just outside Tombstone in his beloved Arizona, in full view of Cochise's Stronghold, near the San Pedro River where twenty years earlier he had first camped, labored, dreamed, and ultimately triumphed, Ed Schieffelin's friends staked his last claim.[11]

---

10. Lockwood, *Pioneer Portraits*, 189; Erwin, *The Southwest of Frank Slaughter*, 182–83; and Faulk, *Tombstone*, 187–88.

11. "Tombstone in Mourning As Prospector-Founder Is Laid to Rest With Simple Rites," *Tombstone Epitaph*, in Green, "Scrapbook," 44; see also *Tucson Arizona Star*, May 18, 1897, and *Tucson (Ariz.) Citizen*, May 22, 1897.

Glossary

assay. To analyze ore to determine the quality and relative value of gold, silver, or other precious metal therein.
Beck. Ed Schieffelin's prized mule and longtime desert companion. According to some, his best friend.
bidarka. A covered canoe made of seal or walrus skin, with one to three holes for people or supplies, used by the Alaskan indigenous peoples.
color. A prospector's term for gold, silver, or other valuable minerals.
cutting pickets. The laborious and boring chore of cutting stakes and fence posts.
excitement. A term used to denote a discovery of mineral wealth that results in a rush of prospectors to a particular region (for example, Salmon River "excitement").
float. Chunks of ore that have been broken off from a major concentration of minerals and carried down into a valley, usually by ancient rivers and streams.
grub. A prospector's term for food. Often it refers to an actual meal.
grubstake. Provisions furnished by one prospector or investor to another on condition that profits of a future discovery be shared. Schieffelin uses the term more generally to refer to pulling together an "outfit," regardless of who pays for it.
outfit. An assemblage of articles, food, and equipment needed by a prospector for an expedition. Schieffelin's outfit usually consisted of two mules, a pick, a shovel, a pan, a bucket, a rocker, a tent, clothing (shirts, trousers, boots, slouch hat), culinary utensils (frying pan, iron pot, tin cups, plates), food (usually flour, bacon, salt, sugar, beans), and a gun.
paydirt. Ore that can be mined profitably. Also, success in extracting ore from a placer deposit.
placer mining. The simplest form of mining. Mineral ore is extracted from the sides of streams or in glacial deposits by using a pan, cradle, or washing rocker.
prospect. An indication of ore or other valuable minerals. Schieffelin often refers to good, rich, or poor prospects.
rich diggins. Gold, silver, or other minerals in deposits that are easy to locate and extract.

GLOSSARY

quartz. A common mineral that often carries gold. Schieffelin and other serious prospectors were in search of gold-bearing quartz that could be pulverized using stamp mills, from which gold could be easily extracted.

stamp mill. A machine that crushes ore to powder from which valuable minerals can be easily extracted.

Bibliography

Archival Collections

Arizona Historical Society, Tucson, Arizona
    Edward Schieffelin Papers (1878–1942), including Schieffelin's manuscript copy of his "History of the Discovery of Tombstone, Arizona as told by the Discoverer, Edward Schieffelin."
    Frederick Brunckow File
    Lockwood Collection
    Richard Gird File
Beinecke Rare Book and Manuscript Library, Yale University
    Schieffelin Family Papers. Series IV. "Fourth Generation."

Miscellaneous Manuscript Materials

Bancroft, Hubert Howe. "Arizona: Her Resources and Future Prospects." Bancroft Library, University of California–Berkeley.

———. "Arizona manuscripts." Microfilm reel 1 PD 1–3, including various writings and transcriptions of oral interviews of Edward Schieffelin. Bancroft Library, University of California–Berkeley.

———. "Edward Schieffelin." Biographic sketch. Bancroft Library, University of California–Berkeley.

———. "Edward Schieffelin's Trip to Alaska, 1882–1883." Manuscript. PK-43. Bancroft Library, University of California–Berkeley.

———. "Edward Schieffelin: The Discoverer of Tombstone." Film PD-13. Reel 11, pp. 1–2. Bancroft Library, University of California–Berkeley.

Brant, George. "Mr. Tombstone." Manuscript photocopy of first eighty pages of uncompleted book, 1859. Author's collection.

Green, Neta Guirado. "Scrapbook." An assortment of newspaper clippings, poems, and other items of interest to Mrs. Green over a fifty-year period, ca. 1883–1964. Author's collection.

Rice, M. M. "The Schieffelin Brothers Ed and Albert: Reminiscences of M. M. Rice." Arizona Historical Society, Tucson.

Schieffelin, Edward. *Memoirs*. Sixteen untitled vignettes. Photocopy of original manuscript, ca. 1959. Author's collection.

Untitled episode. *Death Valley Days*. Radio script. Columbia Broadcasting System. Broadcast February 29, 1944. Author's collection.

Books, Articles, and Dissertations

Allan, Chris. "'On the Edge of Buried Millions': Edward Schieffelin's Search for Gold in the Yukon River, 1882–1883." *Alaska History* 28, no. 1 (2013): 21–39.

Bagley, Will. *Blood of the Prophets: Brigham Young and the Massacre at Mountain Meadows*. Norman: University of Oklahoma Press, 2002.

Bancroft, Herbert Howe. "Life of Richard Gird." In *Chronicles of the Builders of the Commonwealth: Historical Character Story*, vol. 3, 80–90. San Francisco: The History Company, 1891–92.

Barnes, Will C. *Arizona Place Names*. Tucson: University of Arizona Press, 1935.

Beal, Merril D., and Merle W. Wells. *History of Idaho*. 3 vols. New York: Lewis Historical Publishing Company, 1959.

Bartlett, Richard A. *Great Surveys of the American West*. Norman: University of Oklahoma Press, 1962.

Billington, Ray Allen. *America's Frontier Heritage*. Albuquerque: University of New Mexico Press, 1975.

Brandes, Raymond. *Frontier Military Posts of Arizona*. Globe, Ariz.: Dale Stuart King, 1960.

Brooks, Alfred Hulse. *Blazing Alaska's Trails*. University of Alaska and Arctic Institute of North America, Cadwell, Id.: Caxton Printers, 1953.

Brooks, Juanita. *The Mountain Meadows Massacre*. Norman: University of Oklahoma Press, 1962.

———. *John Doyle Lee, Zealot—Pioneer Builder—Scapegoat*. Glendale, Calif.: Arthur H. Clark, 1962.

Breakenridge, William M. *Helldorado: Bringing Law to the Mesquite*. Boston: Houghton Mifflin, 1928.

Burns, Walter Noble. *Tombstone: An Iliad of the Southwest*. New York: Gossett and Dunlap, 1929.

Coutts, R. C. *Yukon: Places and Names*. Sidney British Columbia: Grays Publishing Limited, 1980.

Denton, Sally. *American Massacre: The Tragedy at Mountain Meadows, September 1857*. New York: Knopf, 2003.

Drysdale, A. C. "From Tombstone to the Yukon: Ed Schieffelin's Alaska Expedition," *The Alaska Journal* 13, no. 3 (Summer 1983): 10–15.

Erwin, Allen A. *The Southwest of John H. Slaughter, 1841–1922*. Glendale, Calif.: Arthur H. Clark, 1965.

Ewing, Russell C. "New Light on Chochise," *Arizona and the West* 11, no. 1 (Spring 1969): 57–58.

Faulk, Odie B. *Tombstone: Myth and Reality*. New York: Oxford University Press, 1972.

Ferris, Robert, ed. *Soldier and Brave: Historic Places Associated with Indian Affairs and Indian Wars in the Trans-Mississippi West*. Washington, D.C.: National Park Service/U.S. Department of the Interior, 1971.

Gates, Michael. *Gold at Fortymile Creek: Early Days in the Yukon*. Vancouver: University of British Columbia Press, 1994.

Gird, Richard. "True Story of the Discovery of Tombstone." *Out West* 27 (July 1907): 35–50.

Gruening, Ernest. *The State of Alaska*. New York: Random House, 1968.

Heller, Herbert. *Sourdough Sagas*. Cleveland: World Publishing, 1967.

Hunt, William R. *Alaska: A Bicentennial History*. New York: Norton Company, 1976.

Hulley, Clarence C. *Alaska, 1741–1953*. Portland: Binford and Mort, 1953.

An Illustrated History of Southern California. Chicago: Lewis Publishing, 1890.

Jacobsen, Johan Adrain. *Alaskan Voyage, 1881–1883*. Chicago: University of Chicago Press, 1977.

Knutson, Arthur E. "Steamer New Racket." Six-page unpublished article in Photographic research files (Farciot). Huntington Library. San Marino, California.

Lage, Patricia L. "History of Fort Huachuca, 1877–1913." Master's thesis, University of Arizona, 1949.

Lockwood, Frank. *Arizona Characters: Selected Vignettes*. Los Angeles: Times Mirror Press, 1928.

———. *Pioneer Portraits*. Tucson: University of Arizona Press, 1968.

Limerick, Patricia Nelson, Clyde A. Milner II, and Charles E. Rankin. *Trails: Toward a New Western History*. Lawrence: University Press of Kansas, 1991.

Mathews, Richard. *The Yukon*. New York: Holt, Rinehart and Winston, 1968.

McClintock, James H. *Arizona Prehistoric, Aboriginal, Pioneer, Modern*. Chicago: S.J. Clarke, 1916.

McQuesten, Leroy. *Recollections of Leroy N. McQuesten of Life in the Yukon, 1871–1885*. Dawson City, Yukon Terr.: Yukon Order of Pioneers, June 1952 (reprint 1977).

Mercier, Francois Xavier. *Recollections of the Yukon: Memoirs from the Years 1868–1885*. Anchorage: Alaska Historical Society, 1986.

Paul, Rodman W. *Mining Frontiers of the Far West, 1848–1880*. New York: Holt, Rinehart & Winston, 1963.

Powell, John W. *The Exploration of the Colorado River and Its Canyons*. New York: Dover, 1961 (reprint).

Probert, Thomas. *Lost Mines and Buried Treasures of the West*. Berkeley: University of California Press, 1977.

Rowe, Jeremy. "Arizona Views of Charles O. Farciot." *Journal of Arizona History* 28 (Winter 1987): 373–90.

Rundell, Walter, Jr. "Concepts of the 'Frontier' and the 'West,'" *Arizona and the West* 1, no. 1 (Spring 1959): 13–41.

Schieffelin, Edward Lawrence, and Marilyn Butler, ed. *Destination Tombstone: Adventures of a Prospector*. Mesa, Ariz.: Royal Spectrum Publishing, 1996.

Schieffelin, Edward Lawrence, and Ben T. Trawick, ed. *History of the Discovery of Tombstone, Arizona as told by the Discoverer Edward Lawrence Schieffelin*. Tombstone, Ariz.: Red Maries Books on the West, 1988.

Schwatka, Frederick. *Along Alaska's Great River*. Anchorage: Alaska Northwest Publishing Company, 1885 (reprint, 1983).

Sherwood, Morgan B. *Alaska and Its History*. Seattle: University of Washington Press, 1967.

Smith, Cornelius Cole. *Fort Huachuca: The Story of a Frontier Post*. Washington, D.C.: U.S. Department of the Army, 1981.

Smith, Melvin T. "The Colorado River: Its History in the Lower Canyons Areas." Ph.D. diss., Brigham Young University, 1972.

Thrapp, Dan L. "Dan O'Leary, Arizona Scout: A Vignette." *Arizona and the West* 7, no. 4 (Winter 1965): 287–98.

Turner, Frederick Jackson. *The Frontier in American History*. New York: Henry Holt, 1920.

Turner, William M. "Ed Schieffelin: The Founder of Tombstone, Arizona, and the Story of the Red Blanket Mine." *Newsletter of the Southern Oregon Historical Society* 3, no. 1 (January 1983): 13–18.

Underhill, Lonnie. *The Silver Tombstone of Edward Schieffelin*. Tucson, Ariz.: Roan House Press, 1979.

———. "The Tombstone Discovery: The Recollections of Ed Schieffelin and Richard Gird." *Arizona and the West* 21, no. 1 (Spring 1979): 37–76.

Walker, Ronald W., Richard E. Turley Jr., and Glen M. Leonard. *Massacre at Mountain Meadows*. New York: Oxford University Press, 2008.

Way, W. Jack. *The Tombstone Story*. Tombstone, Ariz.: Tombstone Printers, 1965.

Webb, Melody. *Yukon: The Last Frontier*. Lincoln: University of Nebraska Press, 1985.

Wharton, David B. *The Alaska Gold Rush*. Bloomington: Indiana University Press, 1972.

Wheeler, George M. *Preliminary Report Concerning Exploration and Survey Principally in Nevada and Arizona*. Washington, D.C.: Government Printing Office, 1872.

BIBLIOGRAPHY

Young, Jack. "Jack Young's Adventure." *Alaska Journal* 13, no. 3 (Summer 1983): 16–17.
Young, S. Hall. *Adventures in Alaska*. New York: Fleming H. Revelle Company, 1919.
Young, Otis E. *Western Mining: An Informal Account of Precious Metals Prospecting, Placering, Lode Mining, and Milling on the American Frontier from Spanish Times to 1893*. Norman: University of Oklahoma Press, 1970.

Newspapers

Arizona Daily Citizen (Tucson)
Arizona Daily Star (Tucson)
Denver Tribune (Colorado)
Los Angeles Times (California)
News and Review (Roseburg, Oregon)
Reporter (Portland, Oregon)
Tombstone Epitaph (Arizona)
Tucson Citizen (Arizona)

Interview

Craig, W. Chase. Interview by author, April 22, 1993, Westlake Village, California.

Index

Africa, 8, 69, 98
Alameda, Calif., 9, 99
Alaska, 8, 69, 70–71, 79n24. *See also* prospecting travels: Alaska
Alaska Commercial Company, 74, 79n23, 80, 80n25, 85n34, 88n41, 94; trading posts, 74n14, 75n, 79, 79n24, 81n27, 81n28, 84, 84n33
Alaska expedition, 8, 9, 15, 70–94, 96, 98
Andreafski, Alaska, 79, 79n24, 80
Anna (boat), 36–37. *See also* Grand Canyon tragedy
Arizona, 10, 46, 69, 94. *See also* prospecting travels: Arizona
Austin, Nev., 47

Bancroft, Herbert Howe, 10, 15, 71
Beck (mule), 58, 64, 64n25, 65
Bidwell, T. J., 68n39, 69n44
Boise, Idaho, 26, 34
box canyon flash flood, 46–47, 46n5
Boyer, John "Jack" Oliver. *See* Oliver (John "Jack" Oliver Boyer)
Brown, Mary Elizabeth. *See* Schieffelin, Mary Elizabeth Brown (wife)
Brunckow Mine, 48, 48n8, 53, 58, 64
Burns, George, 24, 25

California, 5, 19, 36n. *See also* prospecting travels: California
Camp Huachuca, Ariz., 7, 51 and n, 57, 59, 66
Camp Walapai, 44, 44n3, 45, 46. *See also* unforgettable experiences: Juniper Canyon ride
Canyonville, Ore., 4, 16, 99, 99n5, 101
Canyonville "excitement," Ore., 4, 101
Cave Valley, Nev., 27
centipede scare, 52
Champion Mine, 62
Charleston, Ariz., 53
Cochise's stronghold, 49, 58, 58n10, 64, 64n26, 65. *See also* Indians: Arizona
Colorado, 6, 10. *See also* prospecting travels: Colorado
Colorado River, 27, 34, 35, 35n4, 42, 44. *See also* Grand Canyon, Ariz.
Contention Ledge Mine, 11, 65n30, 66, 66n33, 68
Cook party (prospectors), 35–37
Corbin family, 66–67, 66n34, 67n37
Corbin Mill and Mining Company, 68n43
Crescent, Ore., 19, 20

Days Creek (South Umpqua River), Ore., 99, 99n5

INDEX

Death Valley, Calif., 6
Defense Mine, 65n31
Desert Act, 53, 53n13
Dulin, Bush, 35–37, 39–42
Dunham, Charlotte Schieffelin (sister), 97

escaped lunatic (Timothy Malloy), 48–51
Eureka, Nev., 47
"excitements" (gold or silver rushes), 11, 48, 56. *See also under individual locations*

Farciot, Charles O. (old "Charlie"), 73, 73n7, 76
flash floods, 46n5. *See also* unforgettable experiences: box canyon flash flood
Fort Rock, Ariz., 44, 45. *See also* unforgettable experiences: Juniper Canyon ride
Fraser River "excitement," British Columbia, 18, 18n3, 70

Gila River, 61
Gird, Richard, 7, 11n, 55, 56, 56n4, 93n46; Tombstone Mine partner, 63–66, 63n24, 66n33, 66n34, 66n35, 67, 67n38, 68n41, 69, 69n44
Globe, Ariz., 60, 61, 62
gold rushes ("excitements"). *See under individual locations*
Goodnow, Charlie, 41, 42
Grand Canyon, Ariz., 6, 27, 34, 35
Grand Canyon tragedy, 35–42, 35n4
Grand Central Mine, 65, 65n30, 65n31, 66n33

Griffith, William T., 48–50, 48n8, 53–54
Guirado, Dr. Ralph C., 14n13, 14n14
Guirado, Elizabeth Jane ("Lizzie") Schieffelin (sister), 14, 14n13, 14n14, 97

H. L. Tiernan (schooner), 8, 73, 73n11, 74n14
Humboldt River, 6, 26

Idaho, 6, 10
Indians, 10, 20, 43; Alaska, 71, 74–75, 78, 80, 81–83, 81n28, 84, 85, 86, 86n39, 87–88; Arizona, 7, 31–33, 40, 43, 44–46, 49 and n, 53, 54, 56–57, 57n7, 58, 59, 60, 61, 67; Nevada, 33; Oregon, 20; Utah, 27, 28–29. *See also* Cochise's stronghold (Ariz.); Mountain Meadows massacre; Virgin Canyon close call
Ivanpah, Calif., 48

Jackson County, Ore., 19, 99
Jacobsen, Johan Adrian, 71, 78n, 79, 79n25
Jewett's Ferry, Ore., 5, 18n1
Juneau, Alaska, 71
Juniper Canyon ride, 44–46

Klondike "excitements," 70

Lady Jennings (boat), 36–37. *See also* Grand Canyon tragedy
La Junta, Colo., 8
Lee, John D., 28, 30
Los Angeles, Calif., 9, 18n1, 24n, 97
Lucky Cuss Mine, 65, 65n29, 66n34

INDEX

mad dogs in Alaska, 81–83, 81n30
Magil, George, 40, 42
Malloy, Timothy (escaped lunatic), 48–51
Maybeso Gulch. *See* Schieffelin Creek (Yukon River)
Mayo, Al, 84, 84n34, 86, 86n39, 91, 93, 93n49
McCracken Mines, 54, 54n14, 60, 60n15, 63
Mineral Park, Ariz., 42
Montana, 10
Montana "excitement," 6, 22
Mormons, 10, 27, 31–32. *See also* Mountain Meadows massacre
Mountain Meadows massacre, 27–30, 30n5, 30n6, 31
Muddy River, 27, 33, 34, 35

Neumann, Henry, 77, 77n20
Nevada, 6, 10, 46, 99. *See also* prospecting travels: Nevada
New Mexico, 6, 36n. *See also* prospecting travels: New Mexico
New Racket (steamer), 8, 73–74, 73n11, 75, 76, 76n16
New York, N.Y., 8, 18, 98
Novikakit, Alaska, 87
Nuklukayet, Alaska, 81, 81n27, 84
Nulato, Alaska, 81, 81n28, 83, 87

old-fashioned country home dinner, 61–62
Old Iritaba district, Ariz., 43
Oliver (John "Jack" Oliver Boyer), 37 and n, 38, 41, 42, 65, 65n30, 65n31, 66n33
Oregon, 10, 14, 16, 20, 23, 48
Owens River, 6, 26
Owl's Nest Mine, 65n29

Palmer (self-styled prospector), 23, 24, 25
Papago expedition, Ariz., 98
Parsons, W. C., 64n27, 66
Peterson, Charlie, 77, 77n19, 78, 79, 79n23
Philadelphia, Penn., 8, 69, 98
Pioche, Nev., 26, 27, 33, 42
Powell, John Wesley, 34
Prescott, Ariz., 43
prospecting travels: Alaska, 70–94, 96, 98; Arizona, 3, 6, 7, 34, 42, 44, 47, 48, 55, 56–69, 98, 102; California, 6, 26, 98, 99; Colorado, 6, 68n41; Idaho, 6; Nevada, 26, 27, 34, 35, 47, 99; New Mexico, 68n41, 96; Oregon, 98–99; Utah, 6, 26

Red Blanket Mine (conjectured), 4, 14, 16, 101
Rogue River, 5, 18n1, 23, 99
Roseburg, Ore., 99

Safford, Anson P. K., 66, 66n36, 67n37, 67n38
Salmon River "excitement," Idaho, 6, 20, 20n4
Salt Lake City, Utah, 8, 26, 96
Salt Lake "excitement," Utah, 26
Salt River "excitement," California, 6
Sampson, W. H., 59, 59n12, 59n13
San Bernardino, Calif., 48, 56
San Francisco, Calif., 9, 19, 98; jump-off for Alaska, 8, 72, 73, 74, 74n13, 83, 93, 94, 96
San Pedro River, 7, 49, 51, 53, 54, 57, 58n11, 61, 63n24, 64, 102
Santa Cruz River, 52
Sauerbrey, (young) Charlie, 73, 73n9, 91

INDEX

Schieffelin, Albert Eugene ("Al") (brother), 6, 9, 12, 24 and n, 47, 60, 60n14, 62, 63; Tombstone Mine partner, 7, 11n, 55, 56, 63, 63n24, 67, 67n38, 68, 68n41, 68n43

Schieffelin, Charles (brother), 102

Schieffelin, Charlotte (later Dunham) (sister), 97

Schieffelin, Clinton Emanuel Del Pela (father), 6, 9, 12, 18n1, 18n2

Schieffelin, Edward Lawrence ("Ed"): birth and genealogy, 5, 18, 18n1, 18n2; civvy street celebrity, 8, 98; death and burial, 14, 16, 99–101; fear of quitting prematurely, 12; malarial fever, 47–48, 65, 94; marriage, 8–9, 98; no more "excitements," 6, 10, 48, 56, 70; passion for prospecting, 8, 9, 47, 98; youth, 5–6, 9, 19–22. *See also* prospecting travels; Tombstone Mine; unforgettable experiences

Schieffelin, Effingham Lawrence ("Eff") (brother), 6, 12, 99; Alaska expedition, 15, 70, 71, 72, 73, 73n10, 84, 86

Schieffelin, Elizabeth Jane ("Lizzie") (later Guirado) (sister), 14, 14n13, 14n14, 97

Schieffelin, Jacob (brother), 99

Schieffelin, Jacob Jr. (grandfather), 18n1

Schieffelin, Jacob Sr. (great-grandfather), 18n1

Schieffelin, Jacob (uncle), 5

Schieffelin, Jane L. Walker (mother), 8, 14n14, 18, 18n2, 102

Schieffelin, Mary Elizabeth Brown (wife), 8–9, 102

Schieffelin Creek (Yukon River), 3, 85, 85n37

Schieffelin Hall, 3 and n, 102

Shamrock (boat), 36–40. *See also* Grand Canyon tragedy

Signal Mine, 63, 63n21, 64

Silver City, Idaho, 47

Silver King Mine, 60, 60n14

silver rushes ("excitements"). *See under individual locations*

Smith, Albert ("Alvah"), 48–50, 48n8, 53, 58n9, 64n27, 66

Snake River, 26, 37

South Dakota, 10

St. Michael, Alaska, 75n, 78, 83, 86, 88, 89, 94; jump-off for Alaskan interior, 8, 73n11, 74, 75, 77

Stonewall Jackson Mine, 60–61, 61n16

St. Thomas, Nev., 33, 34, 35, 35n4, 38, 42

Tioga County, Pa., 5, 18, 19

Tombstone, Ariz., 14, 52, 65, 69, 101–102

Tombstone "excitement," Ariz., 54, 66, 66n32

Tombstone Gold and Silver Mill and Mining Company, 7, 67n38, 68n43

Tombstone Mill and Mining Company (amalgamation), 68n43

Tombstone Mine, 3, 11 and n, 16, 47, 48, 52, 66n34, 98, 101; discovery of, 7, 55–68, 68n39; how it was named, 7, 55, 57, 57n8

Tombstone mining district, 7, 52, 53, 55, 56, 61

Tough Nut Mine, 66n34

Tozer, Charles M., 66, 66n34

Tucson, Ariz., 7, 52, 56, 59, 67

Turner, Frederick Jackson, 4

Unalaska, Alaska, 74, 74n14, 77
unforgettable experiences: box canyon flash flood, 46–47, 46n5; centipede scare, 52; escaped lunatic (Timothy Malloy), 48–51; Grand Canyon tragedy, 35–42, 35n4; Juniper Canyon ride, 44–46; mad dogs in Alaska, 81–83, 81n30; old-fashioned country home dinner, 61–62; Virgin Canyon close call, 31–33
Utah, 6, 27, 31. *See also* Mountain Meadows massacre; prospecting travels: Utah

Virgin Canyon close call, 31–33
Virgin River, 31, 33
Vosburg, John, 66n35, 66n36, 67n38

Walapai mining district, 56
Walker, Jane L. (mother). *See* Schieffelin, Jane L. Walker (mother)

Walker, Joseph (uncle), 18
Warren, Charlie, 99, 101
Washington (D.C.), 8, 98
Washington (state), 10
Western Fur Trading Company, 74n14, 81n28, 89
Westside Mine, 65, 65n28
Wheeler, George M., 34
White, Josiah H., 64n27, 66, 66n33
Williams, Henry "Hank" D., 65, 65n30, 65n31, 66n33
Winnemucca, Nev., 26, 47
Woodville, Ore., 99
Woolfe, H. D., 71, 79, 80n25, 81

Young, Jack, 73, 73n8, 78, 86n39, 92
Yukon (steamer), 80, 80n26, 81, 84n34, 94n49
Yukon River, 3, 8, 10, 70, 72, 76, 78, 85, 85n37, 89, 91, 93n48
Yukon Territory, 3, 8

www.ingramcontent.com/pod-product-compliance
Lightning Source LLC
Chambersburg PA
CBHW020857160426
43192CB00007B/961